T0205504

Techniques in Life Science and Biomedicine for the Non-Expert

Series Editor

Alexander E. Kalyuzhny, University of Minnesota
Minneapolis, MN, USA

The goal of this series is to provide concise but thorough introductory guides to various scientific techniques, aimed at both the non-expert researcher and novice scientist. Each book will highlight the advantages and limitations of the technique being covered, identify the experiments to which the technique is best suited, and include numerous figures to help better illustrate and explain the technique to the reader. Currently, there is an abundance of books and journals offering various scientific techniques to experts, but these resources, written in technical scientific jargon, can be difficult for the non-expert, whether an experienced scientist from a different discipline or a new researcher, to understand and follow. These techniques, however, may in fact be quite useful to the non-expert due to the interdisciplinary nature of numerous disciplines, and the lack of sufficient comprehensible guides to such techniques can and does slow down research and lead to employing inadequate techniques, resulting in inaccurate data. This series sets out to fill the gap in this much needed scientific resource.

More information about this series at https://link.springer.com/bookseries/13601

Nayana Patil • Aruna Sivaram

A Complete Guide to Gene Cloning: From Basic to Advanced

Nayana Patil
School of Bioengineering Sciences
& Research
MIT ADT University
Pune, Maharashtra, India

Aruna Sivaram
Associate Professor, School of
Bioengineering Sciences & Research
MIT ADT University
Pune, Maharashtra, India

ISSN 2367-1114 ISSN 2367-1122 (electronic)
Techniques in Life Science and Biomedicine for the Non-Expert
ISBN 978-3-030-96853-3 ISBN 978-3-030-96851-9 (eBook)
https://doi.org/10.1007/978-3-030-96851-9

This Springer imprint is published by the registered company Springer Nature Switzerland AG
The registered company address is: Gewerbestrasse 11, 6330 Cham, Switzerland

Foreword

I would like to appreciate the efforts of the authors in bringing out the book in its current form. With recombinant DNA technology and synthetic biology emerging as leading technologies in addressing several global challenges, this book provides a timely intervention in initiating young students to the field. The book begins with an introduction into the history of rDNA technology and then delves into the technical details. It emphasizes on the technical aspects of vectors, gene isolation methods, cloning strategies, and the screening strategies of transformants in significant detail. Advanced gene editing techniques like zinc finger nucleases, transcription activators like effectors, RNA interference, CRISPR Cas system, Cre-lox and their application in various fields including healthcare, agriculture, and environment sustainability have been elaborated. A detailed discussion on these recent trends and their applications will kindle young minds with new ideas and inculcate research aptitude early on in life. The final chapter on bioethics and safety is an important one for the student audience. Another notable approach of this book is the clear illustrations used to explain different topics in an attractive and informative manner. The tables in each chapter collate several information available in literature and give a concise view of the topic. Every chapter concludes with a crisp summary and a few conceptual questions which ignite the reader's mind. Besides Dr. Nayana and Dr. Aruna from MIT School of Bioengineering Sciences and Research and several authors from academia, research, and industry have contributed towards this book. I wish them all a great success in the reception of the book.

MIT School of Bioengineering Sciences and Research
MIT ADT University
Pune, India

Renu Vyas

Preface

Anyone who has studied even the basics of biology would recognize and appreciate the importance of cloning technologies and its contribution to the progress of this science. Cloning or recombinant DNA technology is perhaps one of the most applied techniques in Biology. Through this book, we aim to provide concise information to beginners, thereby generating their interest and helping them develop a keen acumen in cloning techniques. This will be a one of a kind book written in lucid language, devoid of scientific jargons. If at all these jargon have to be used, they would be clearly explained in simple terms.

We begin the book by giving a quick peep into the history of cloning techniques by discussing all major milestones in cloning/recombinant DNA technology. This is followed by more technical concepts like the cloning strategies, the inserts, the vectors, the restriction enzymes, and the DNA transfer protocols. All basic aspects pertaining to cloning in prokaryotic and eukaryotic cells along with the screening strategies have been covered concisely, but with enough details and illustrations to make it interesting, attractive, and informative.

We conclude the book by deliberating on the ethical and safety aspects of recombinant DNA technology. This chapter, we presume is important especially given that we are addressing readers who are either beginners or novices in the field. Our vision is not only to create resources with a sound knowledge of the techniques, but also to develop a sense of responsibility and accountability in them.

Every chapter in this book has been followed by a brief summary where we discuss all salient points covered. The book is rich in illustrations. Each chapter has question and answer section, which the readers can use as samples to review their understanding of the topics.

We hope this book would be a useful guide to people to develop a basic understanding of cloning/recombinant DNA technologies.

Pune, India
E-mail: nayana.patil@mituniversity.edu.in
Nayana Patil, Ph.D.

Pune, India
E-mail: aruna.sivaram@mituniversity.edu.in
Aruna Sivaram, Ph.D.

Acknowledgement

We would like to express our heartfelt gratitude to our families for putting up with our erratic work schedule. But for the support from our spouses, children, and parents, we would have not been able to burn the midnight lamp, and this book would have remained a distant dream.

We also take this opportunity to thank our Director, Prof. Vinayak Ghaisas and management of MIT ADT University, Pune, India for their support. Our sincere gratitude to our Head of the School, Dr. Renu Vyas for the extraordinary encouragement that we received from her. We got inspired from her dedication, time management and multi-tasking capabilities which energized us to complete this work successfully.

A very big thanks to all our contributors Dr. Manisha Modak, Dr. Narendra Nyayanit, Dr. Shalmali Bivalkar, Dr. Rajeev Mehla, Dr. Rajendra Patil, Dr. Sangeeta Sathaye, Dr. Neeraj Maheshwari, Mr. Praveen Kumar, Dr. Shriram Rajpathak, Ms. Rupali Vyawahare, Dr. Madhura Chandrashekar, Dr. Manjula Maralappanavar, Dr. Premjyoti Patil, Dr. Shruti Desai, Dr. Madhura Vipra, and Dr. Richa Sharma for their genuine efforts. We appreciate and acknowledge all the hard work that you have engaged in making this work a reality, despite your busy schedule. We would also like to thank Ms. Madhavi Bhogale for her help in preparation of the manuscripts.

Dr. Alexander Kalyuzhny, the editor of the series deserves special mention here. We would like to thank him for giving us this opportunity. Ms. Alison Ball and her team from Springer Nature coordinated with us in an excellent way. Thanks to Alison and her team for making this a success.

We fall short of words to thank our alma matter, our teachers, research guides, colleagues, and friends. We have learnt quite a lot from you and have drawn motivation from each one of you; that has made us what we are today. Your wishes, guidance and blessings were the key elements which helped in this work seeing the light of the day.

Contents

Contributors

Shalmali Bivalkar National Centre for Cell Science, Pune, Maharashtra, India

Madhura Chandrashekar School of Bioengineering Sciences & Research, MIT ADT University, Pune, Maharashtra, India

Shruti Desai Department of Pathology, Yale University School of Medicine, New Haven, CT, USA

Praveen Kumar Kashiv Biosciences, Ahmedabad, Gujarat, India

Neeraj Maheshwari School of Bioengineering Sciences & Research, MIT ADT University, Pune, Maharashtra, India

M. S. Maralappanavar University of Agricultural Sciences, Dharwad, Karnataka, India

Rajeev Mehla Serum Institute of India, Pune, Maharashtra, India

Manisha Modak Department of Zoology, Sir Parashurambhau College, Pune, Maharashtra, India

Narendra Nyayanit Department of Zoology, Sir Parashurambhau College, Pune, Maharashtra, India

Premjyoti Patil Basaveshwar Engineering College, Bagalkote, Karnataka, India

Rajendra Patil Department of Biotechnology, Savitribai Phule Pune University, Pune, Maharashtra, India

Shriram Rajpathak ChAdOx1 and COVOVAX Vaccine Department, Serum Institute of India Pvt Ltd., Pune, Maharashtra, India

Sangeeta Sathaye Department of Biotechnology, Modern College of Arts Science and Commerce, Shivajinagar, Pune, Maharashtra, India

Richa Sharma Labcorp Scientific Services & Solutions Pvt. Ltd, Pune, Maharashtra, India

Aruna Sivaram School of Bioengineering Sciences & Research, MIT ADT University, Pune, Maharashtra, India

Madhura Vipra Medvolt Tech Pvt. Ltd, Pune, Maharashtra, India

Rupali Vyawahare SGS Canada, Mississauga, ON, Canada

About the Contributors

Shalmali Bivalkar is currently working as Junior Scientist at Vaccine Testing Facility, National Centre of Cell Science (NCCS), Pune, India. She received her Ph.D. in Biomedical Sciences from the University of South Carolina, USA. Her postdoctoral research at the University of Pittsburgh, USA, involved work on HIV-1 neuropathogenesis. She has successfully completed a 3-year DST-WOSA fellowship with funding from the Department of Science and Technology at NCCS, Pune. Her fields of expertise include virology, molecular biology, recombinant DNA technology, and cell biology.

Madhura Chandrashekar is currently working as an Assistant Professor at the School of Bioengineering Sciences and Research, a constituent unit of MIT ADT University, Pune. She obtained her doctorate from the University of Agricultural Sciences, Dharwad. During PhD she was involved in the development of transgenic cotton. She has more than 10 years of research and teaching experience. Her current research interest includes bioremediation and plant molecular biology.

Shruti S. Desai, Ph.D., is an Associate Research Scientist at the Department of Pathology, Yale University, CT, USA. She received her Ph.D. from the University of Pune. After working at Abgenics Lifesciences in Pune, she did her postdoctoral training at the National Institutes of Health (NIH), Maryland, and Yale University. The focus of her work is discovery of biomarkers in cancer and study their relevance with T cell exhaustion and association with clinical benefit.

Neeraj O. Maheshwari is currently working as Assistant Professor at the School of Bioengineering Sciences & Research, MIT ADT University, Pune, India. He obtained her Ph.D. from Université de Technologie de Compiègne, France, in Biomaterials and Bioengineering. He has research experience as a research fellow at the Institute of Bioinformatics and Biotechnology, SPPU, Pune, and Nanoscience unit IISER-Pune. His research interest comprises biomaterials (micro and nano) for drug delivery, diagnostics, and imaging. He also holds interest in using biomaterials as scaffolds for tissue engineering and biomedical research.

Manjula S. Maralappanavar is currently working as a senior cotton breeder at Agricultural Research Station (cotton), University of Agricultural Sciences, Dharwad. She has experience in cotton breeding and biotechnology for more than two decades from 1997 to date at UAS, Dharwad. She has developed a genotype-independent transformation protocol in cotton. UASD Transgenic Cry1Ac Cotton Event No. 78 has been developed which is superior to the commercially cultivated event. The event is presently under biosafety research field trial, and efforts are in progress for pyramiding with Cry2ab and Vip genes independently.

Rajeev Mehla is currently working as Section Head of COVID and Dengue Vaccine R&D and QC at Serum Institute of India, Pune, India. He obtained his Ph.D. from the National Institute of Virology, Pune, India. His postdoctoral research involved molecular pathogenesis and neuroinflammation of HIV and antivirals at the University of South Carolina, USA, and the University of Pittsburgh, USA. He has a rich experience in both industry and R&D in the field of viral vaccines, virology, molecular biology, and recombinant DNA technology.

Manisha Modak is currently working as Assistant Professor at the Department of Zoology, Sir Parashurambhau College, Pune, India. She obtained her Ph.D. from the Department of Zoology, Savitribai Phule Pune University, Pune. She had completed a couple of projects funded by UGC and BCUD and also has a number of publications to her credit. Her research interests include oxidative stress, DNA damage and repair, diabetes, and drug discovery.

Nayana Patil, Ph.D., is currently working as Associate Professor at the School of Bioengineering Sciences & Research, MIT ADT University, Pune, India. She obtained her Ph.D. from the University of Pune, India. Her postdoctorate research work involved identification of protein biomarkers and correlation to epigenetics in mouse and mammalian cell lines as a model system. She has over 9 years of teaching experience in biotechnology. Her areas of expertise include microbiology, synthetic biology, proteomics, and nano-biotechnology.

Premjyoti Patil is currently working as an Assistant professor at Basaveshwar Engineering College, Department of Biotechnology, Bagalkote, Karnataka, India. She obtained her doctorate from Vishvesvaraya Technological University, Belgaum, India. She has 15 years of teaching experience for undergraduates. Her research interests include algal biotechnology, nutraceuticals, and bioinformatics.

Rajendra Patil, Ph.D., is currently working as Assistant Professor at the Department of Biotechnology, Savitribai Phule Pune University. He obtained his Ph.D. in Biotechnology from the Department of Biotechnology, Savitribai Phule Pune University. His areas of interest include nanoparticle biosynthesis, proteomics, and nano-materials for therapeutics usage.

A.K. Praveen Kumar is currently working as Senior manager of Bioassay department at Kashiv Biosciences Pvt. Ltd., Ahmedabad. He obtained his postgraduation from Bharathiyar University, Coimbatore. Praveen has a rich industry experience in the area of small molecule drug discovery as well as bio-therapeutics. He has worked in Lupin Research Park as senior scientist in the area of immunology and immuno-oncology, prior to which he was associated with Connexios LifeSciences Pvt. Ltd., Bangalore, in the area of metabolic disorders and Magene LifeSciences Pvt. Ltd. in the area of PK-PD analysis.

Shriram Rajpathak, Ph.D., is currently working as Team Manager at Serum Institute of India Limited, Pune. He obtained his Ph.D. from the Department of Zoology, Savitribai Phule Pune University. His research expertise includes epigenetic studies, metagenomics, and DNA methylation.

Sangeeta Sathaye is currently working as an Assistant Professor at Modern College of Arts, Science and Commerce, Shivajinagar, Pune, India. She obtained her Ph.D. in Biochemistry from National Chemical Laboratory, Pune. She has over 10 years of teaching experience in biotechnology. Her areas of interest are protein biochemistry and plant biotechnology.

Richa Sharma is currently working as a Global Feasibility Lead with the Clinical FSPx team at Labcorp Scientific Services & Solutions Pvt. Ltd. Her past experience includes working as a Scientific and Medical Writer for AdisInsight (SpringerNature). She has completed her doctoral studies pertaining to drug development against filariasis, an infective disease, from the RTM Nagpur University, and has several national and international publications and presentations from her research work.

Aruna Sivaram, Ph.D., is currently working as Associate Professor at the School of Bioengineering Sciences & Research, MIT ADT University, Pune, India. She obtained her Ph.D. from the National Institute of Virology, Pune. She has a rich industry experience as a senior research scientist from Lupin Research Park, a top-notch pharmaceutical industry. Her research interests include molecular pathophysiology, synthetic biology, vector biology, and drug discovery.

Madhura Vipra is one of the pioneers set out to revolutionize the healthcare sector of India. Dr Madhura has a M.Sc. in Microbiology with a Ph.D. in Cancer Biology from Biotechnology department of SP Pune University. She is the cofounder and CEO of Medvolt Tech Pvt. Ltd., a unique deep-tech start-up that amalgamates the essential tenets of artificial intelligence and data science with biological sciences.

Rupali Vyawahare is currently working as Biotech Analyst at SGS Canada Inc., Mississauga, Ontario, Canada. She obtained her M.Sc. in Agricultural Biotechnology from Vasantrao Naik Marathwada Agricultural University, Parbhani, India. She has extensive industry experience as an Assay development scientist from Pharma Giants such as Merck & Co. and Pfizer from the USA, Lupin Ltd. India, and SGS Canada Inc. from Canada. Her research interests include cell biology, molecular biology, and drug discovery.

Chapter 1
The Recombinant DNA Technology Era

Manisha Modak, Narendra Nyayanit, Aruna Sivaram, and Nayana Patil

1.1 Introduction

Hungarian engineer Karl Ereky used the word Biotechnology for the first time in 1917. According to him biotechnology was "all lines of work by which products are produced from raw materials with the aid of living things." With continued scientific and technical advancements, biotechnology was redefined as "industrial production of goods and services by processes using biological organisms, systems, and processes" [1]. The key aspect of industrial production is to identify the microorganisms which produce useful products. Initially, the search was focused on finding the microorganisms from different natural sources. But there was a limitation of optimum production of desired product. Even though the strain improvement method was extremely successful, it was tedious, time consuming, costly, and very random in terms of generating the strains. At the same time, one cannot predict the effect of mutation on other metabolic processes of the organism. This scenario changed by the emergence of recombinant DNA (rDNA) technology which enabled the manipulation of the genetic material at the desired positions. The definition of rDNA is any artificially created DNA molecule which brings together the genetic sequences that are not usually found together in nature [2]. The foundation stone of recombinant DNA technology (RDT) was set in a lunch time conversation between two scientists Stanley Cohen of Stanford University, and Herbert Boyer of the University of California at a conference in 1973. Cohen was in search of methods to introduce

M. Modak · N. Nyayanit
Department of Zoology, Sir Parashurambhau College, Pune, Maharashtra, India

A. Sivaram · N. Patil (✉)
School of Bioengineering Sciences & Research, MIT ADT University,
Pune, Maharashtra, India
e-mail: aruna.sivaram@mituniversity.edu.in; nayana.patil@mituniversity.edu.in

N. Patil, A. Sivaram, *A Complete Guide to Gene Cloning: From Basic to Advanced*, Techniques in Life Science and Biomedicine for the Non-Expert,
https://doi.org/10.1007/978-3-030-96851-9_1

small circular DNA molecules into bacterial cells and Boyer was trying to cut DNA at specific nucleotide sequences using enzymes. During this conversation both of them realized the beauty of the combination of both the techniques. This resulted in a milestone discovery "Construction of Biologically Functional Bacterial Plasmids *In Vitro*" [3]. Even though these two laid the foundation of the recombinant biotechnology era it was built on the shoulders of many scientists which can be traced back to Gregor Mendel, the father of genetics.

1.2 The Era of Traditional Genetics

The concept of heredity and variation was explained by Gregor Mendel in 1866 on the paper titled "Experiments in plant hybridization." He postulated the principle of heredity based on his experiments with garden pea plants [4]. Mendel had used the term "factor" to denote the physical entity which is responsible for heredity. Many decades later, Boveri observed the individual chromosomes while studying sea urchins and reported that proper embryonic development does not occur unless chromosomes are present. In 1903, he studied mitosis in *Ascaris megalocephala*, a parasitic nematode and demonstrated the continuity of chromosomes during cell division. During the same time, Sutton who was working on grasshopper (*Brachystola magna*) spermatogenesis observed that chromosomes segregate in the process of cell division and gamete formation which was similar to Mendel's findings [5]. These results established the role of chromosomes in heredity and later recognized as Boveri and Sutton's chromosome theory of inheritance. This theory explained two basic characteristics of genetic material—individuality and continuity. The term "gene" was coined by Danish botanist Wilhelm Johannsen in 1909 with a mere abstract concept in mind to describe the fundamental physical and functional units of heredity. Suttons' chromosomal theory of inheritance was experimentally supported by T. H. Morgan and his colleagues. In 1910, Morgan concluded that genes are real, physical objects, located on chromosomes, with properties that can be manipulated and studied experimentally [6]. He also showed that genes were present on chromosomes in a linear order and were linked together [7]. In spite of such classical and breakthrough studies the exact molecular nature of genes was unclear. It was in 1944, through the experiments of Avery, MacLeod, and McCarty that the deoxyribonucleic acid (DNA) was discovered to be the genetic material. Hershey and Chase in 1952 had given the confirmation that DNA is genetic material with their classical experiment with ^{32}P labeled DNA and ^{35}S labeled protein. This gave the simple definition of a gene as a segment of DNA in chromosomes. Later Watsons and Crick's double helical structure of DNA revealed the DNA theory of inheritance. These studies threw more light on genes as the structural unit of heredity. Simultaneously, complementary research programs were going on to unravel the functional entity of the gene. Archibald Garrod in 1902 provided the first evidence for the gene to protein concept by studying the metabolic disorder alkaptonuria which is a result of failure of a specific enzyme [8]. The concept of one gene one enzyme was established by Beadle and Tatum in 1942 while studying X-ray induced mutagenesis in *Neurospora crassa*. They observed three nutritional mutants and identified that they differ from wild type [9]. In order to find out whether mutations belong to same gene or not, complementation based cis trans test was used by Lewis and Benzer who proposed the term cistron to define the genetic unit of function. [10, 11]. The relationship between the structural entity that consists of the nucleotide sequence and the functional entity consisting of the polypeptide sequence was established through the collinear hypothesis by Gamow [12]. Crawford and Yanofsky established the concept of one cistron, one polypeptide [13]. Major landmark discoveries associated with basic molecular biology are mentioned in Table 1.1.

Table 1.1 Milestones in traditional genetics

Year	Name of the scientist	Discovery
1866	Gregor Mendel	Described laws of heredity with the help of "factor"
1869	Frederick Miescher	Isolated DNA from cells for the first time and calls it "nuclein"
1900	DeVries, Correns, and von Tschermak	Independently rediscovered Mendel's work
1902	Boveri	Observed the individual chromosomes
1902	Archibald Garrod	Orderly inheritance of disease
1903	Boveri and Sutton	Chromosomal theory of inheritance
1909	Wilhelm Johannsen	Coined the term "gene"
1910	T H Morgan	Genes are present on chromosome
1941	Beadle and Tatum	1 gene = 1 enzyme
1944	Avery, MacLeod, and McCarty	DNA transferred from generation to generation
1952	Hershey and Chase	Confirmation of genetic material
1953	Francis H. Crick and James D. Watson	Structure of DNA
1955	Arthur Kornberg and colleagues	Isolated DNA polymerase
1951	Lewis and Benzer	Proposed the term cistron
1952	Dounce	Collinear hypothesis—One gene one polypeptide
1957	Francis Crick	Proposed central dogma of molecular biology
1958	Crawford and Yanofsky	One cistron one polypeptide
1958	Matthew Meselson and Franklin Stahl	Semiconservative replication of DNA
1966	Marshall Nirenberg and others	Genetic code cracked
1970	Kelly and Smith	Discovery of restriction endonucleases
1971	Danna and Nathans	
1972	Berg and his students	Constructed first rDNA
1973	Boyer, Cohen & Chang	Transformed *E. coli* with recombinant plasmid
1974	Morrow	First animal gene cloned
1983	Kary Mullis	Invention of polymerase chain reaction (PCR)
1997	Ian Wilmut	Cloning an adult animal, Dolly

1.3 The Era of Recombinant DNA Technology

Studies till 1960 evolved the abstract concept of gene to a more meaningful idea that the gene is a segment of DNA and is the structural and functional unit of heredity. Most of the studies which were carried to understand the structure of gene and its regulation till then were in prokaryotes and the viruses which infect them. Similar studies in eukaryotic cells necessitated the development of new methods. One such method which was employed was transduction i.e., transfer of genes from one strain to another strain of bacteria by a bacteriophage. Generalized and specialized transduction experiments were routinely done using phages P1 and lambda (λ), respectively. Scientists thought that a similar gene transfer strategy could be used in mammalian cells using polyoma and SV40 viruses. It was already known that both these viruses could produce viral particles with exclusively mammalian DNA. Inability to pack the entire single mammalian gene and lack of specific cutting and selection methods were the major limitations of this technique. This led to the invention of alternative methods to combine two segments of DNA in vitro. Joining two DNA molecules having no compatible ends was the major hurdle in this process. This problem was solved by terminal deoxynucleotidyl transferase (TdT), an enzyme which adds a single nucleotide tail at the 3′-ends of duplex DNA. This characteristic of TdT was explored by scientists to develop the terminal transferase tailing method to join two DNA molecules in vitro [14].

In 1971 Berg et al. developed a DNA molecule *λdvgal 120* which can be used to link SV40 DNA. λdvgal 120 comprises the genes required for autonomous replication in *E. coli* and an intact gal operon from phage λ. Mertz et al., showed that *λdvgal 120* can be transformed in *E. coli* as a linear molecule and it circularizes to form an autonomously replicating plasmid [15]. Discovery of restriction endonucleases, the enzymes which cleaves DNA at specific site by Kelly and Smith (1970) and Danna and Nathans (1971) was one of breakthrough RDT. With this information in hand Jakson and colleagues synthesized the first rDNA molecule by combining *λdvgal 120* DNA and SV40 DNA using six different enzymes very strategically [16]. A schematic representation of the strategy used for first rDNA is given in Fig. 1.1.

However, concerns were raised regarding safety of this technique due to use of SV40 and *E. coli* The concerns we discussed in Asilomar Conference Center 1973. The conference concluded that the rDNA research should proceed but under strict guidelines.

The experiments continued with strict regulations keeping the benefit of mankind as a primary goal. The discovery that the cleavage with restriction endonuclease *EcoRI* generates cohesive ends reduced the use of multiple enzymes [15]. In 1973, Cohen et al. created pSC101 with a segment of DNA having antibiotic resistance gene from another *E. coli* plasmid. The new plasmid replicated in *E. coli* and also showed antibiotic resistance [3]. This led to the construction of vectors which created wonders in the RDT era. Cohen and his team also constructed a chimeric plasmid from *E. coli* Ps101 and a plasmid from *Staphylococcus aureus*. This interspecies plasmid propagated well in *E. coli* expressing the characters from both the parental plasmids [17]. The cloning of eukaryotic DNA was demonstrated for the first time by ligating ribosomal DNA of *X. laevis* and pSC101, both cleaved with

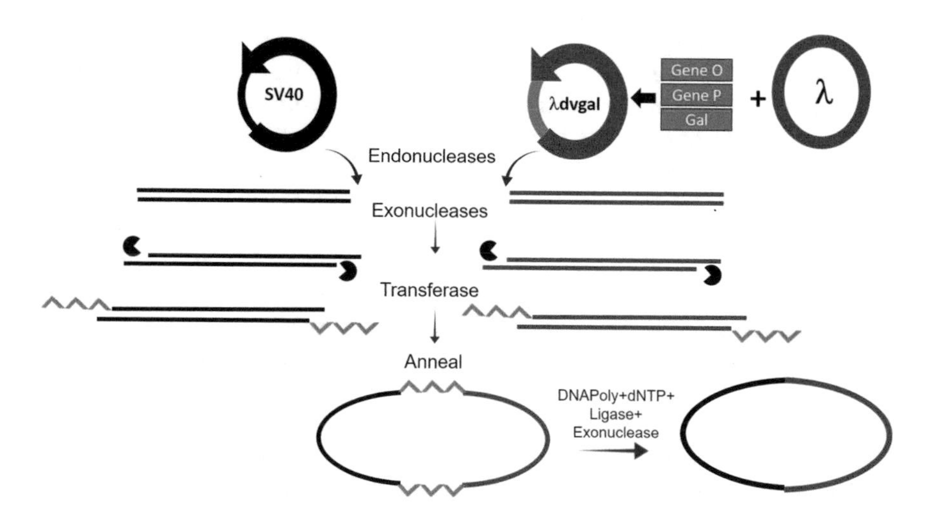

Fig. 1.1 Strategy used by scientist to prepare first rDNA molecule: The λdvgal 120 is a variant of lambda phage that includes three galactose utilization genes from *E. coli* into *λ* phage. The circular SV40 and λdvgal 120 were linearized using a set of endo and exonuclease enzymes. In the next step poly A tail was attached 3′ to SV40 while a poly T tail was attached to 3′ end of λdvgal 120 which allowed hybridization of between the DNA which was followed by covalently joining the DNA using ligase

EcoRI [18]. This experiment proved that DNA from any organism could be cloned and proliferated in *E. coli*. Immediately after the discovery of the first rDNA, Stanford University and the University of California jointly filed a United States patent application citing Cohen and Boyer as the sole inventors of this technology. These claims were approved by the U. S. Patent Office in 1980 with slight modifications and generated a revenue of $300 million.

All these experiments collectively paved the path for rapid production of gene products in large scale even from more complex organisms. Soon the feasibility of using bacteria for cloning human genetic information and its potential application for fighting diseases was realized. This laid the foundation of the first start-up biotech company Genentech in 1976, founded by Boyer and Swanson with a goal to produce new vital medicine by manipulating the genes for commercial purpose. They collaborated with a group of scientists involved in chemically synthesizing DNA to create full-length artificial genes. They selected somatostatin, a small hormone with only 14 amino acids to give the first "proof-of-concept" to the investors. After a few unsuccessful attempts the researchers managed to chemically synthesize and clone human somatostatin in *E. coli* using a plasmid vector. This experiment confirmed that RDT could be used to make human proteins and it opened doors toward unlimited scientific opportunities. This was followed by cloning of the first marketable protein, the human insulin in 1979. Eli Lilly and Company started a joint-venture with Genentech for the development and production of recombinant human insulin which was named Humulin. In 1982 the first biotechnology product Humulin was approved and launched in the market by FDA. This was the milestone in the era of biotechnology and led to some landmark discoveries later.

Further boost to RDT came with the development of novel vectors. From the very early vector *λdvgal 120* till today, there has been a remarkable advancement in vector design and molecular cloning strategies. Initial vectors had a single origin of replication and therefore could be used only in similar types of organisms. A series of vectors called "shuttle vectors" that contain two different origins and two different selection markers were developed soon. These vectors can be used to transfer genes to two distinct organisms. As development continued, reporter genes and multiple cloning sites were added to vectors for selection of positive transformants and for cloning of the desired DNA, respectively. Advancements in vectors and cloning strategies led to the emergence of an entirely new branch called synthetic biology which combines biological parts and modules to create more reliable and robust systems. Further, the invention of Polymerase chain reaction (PCR) in 1983 by Kary Mullis had an impact on RDT. A very small amount of the desired DNA can be amplified using proper primers and enzymes through this technique. PCR conquered all branches of biology within a short period of time due to its sensitivity and robustness. Table 1.2 gives an insight into the major landmarks associated with RDT.

Table 1.2 Milestones in RDT

Year	Achievement
1973	Asilomar conference Center in Pacific Grove, California, concluded the rDNA research should proceed but under strict guidelines
1976	First genetic engineering company
1980	SCOTUS v. Chakrabarty, first patent of superbug
1981	First transgenic mice and fruit flies. First GMO plant
1982	First biovaccine
1982	First recombinant human insulin
1983	Artificial chromosome
1983	Huntington's disease genetic marker discovered
1986	First recombinant vaccine (vaccine for hepatitis B) for humans, approved; first biotech-based anticancer drug (interferon) produced
1988	*Bt* corn and rennin produced from GMO
1990	Immune disorder treated with gene therapy
1994	FDA approval for FLAVR SAVR tomato
1996	Commercialization of first GMO crop
2001	First FDA approved gene-targeted drug Gleevec®, for chronic myeloid leukemia
2002	First time cloning of an endangered species
2006:	First FDA approved recombinant vaccine Gardasil®, against human papillomavirus (HPV)
2009	Recombinant human antithrombin approved
2013	Discovery of CRISPR for genome editing
2014	Reconstruction of artificial functional yeast chromosome
2015	First genetically modified salmon food approval
2016	For the first time scientists used CRISPR to treat illness in human patient
2016	Successful 3D-printing of "heart on a chip"

1.4 Application Array of Recombinant DNA Technology

Continuous advancements in the field of RDT have increased its applications multifold. Beginning with insulin production in *E. coli* many therapeutically useful proteins were made in bacteria. As the technology evolved, genetic manipulation extended to plants and animals which broadened the horizons of the biotechnology industry [19]. It created wide scope for inventions and to produce more beneficial products. Various applications of RDT are mentioned in Fig. 1.2. RDT replaced the conventional strategies and gave a new approach to deal with various issues like malnutrition, diseases, environmental pollution, agriculture-related problems, and many others. Onset of the genomic era shifted the focus of treatment from symptoms to cause. The identification of correlation between phenotype of a disease and genotype of the patients opened several avenues for development of novel approaches in therapeutics. This includes design of small molecule and biologic drugs, developing the animal models with the same genetic defects as seen in patients (disease modeling), correction of the disease using gene therapy, xenotransplantation, etc.

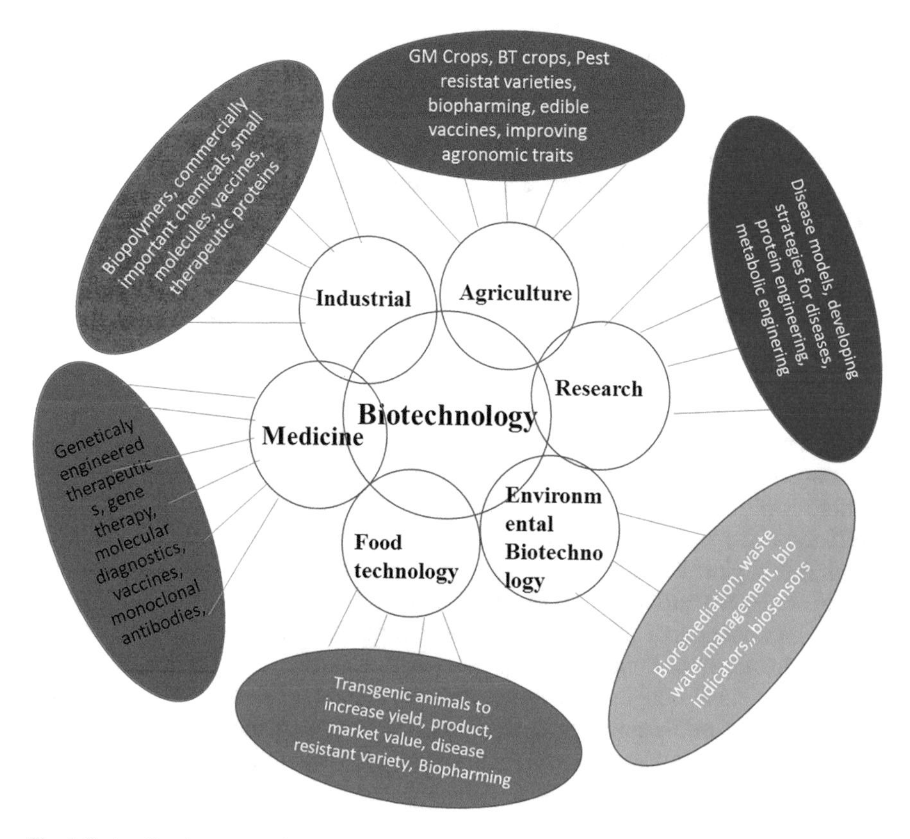

Fig. 1.2 Application array of RDT

RDT also helped in developing new methods for vaccine development like subunit vaccine, developing genetically modified, less pathogenic and more immunogenic vaccines, use of known safe vaccine strains as vectors, plants as edible source of vaccine and DNA vaccine. Metabolic engineering is one of the beneficial applications of RDT where wild-type strains can be converted to strains with enhanced production by making desired changes in a rational way. Many products prepared by RDT have been approved for commercial use. High cost of production of biopharmaceuticals and the time required from research to market approval forced biotechnology companies to search alternatives like biosimilars. Biosimilars are "highly similar" to the original biological drug. There are some minor differences in active ingredients, but no significant change in clinical properties. Recombinant proteins with properties better than the existing drugs have also been synthesized. Such improved proteins are referred to as biobetters. Trastuzumab biobetter, filgrastim biobetter, etc. have been approved by FDA [20].

1.5 Era of Genetically Modified Organisms

Scientists have been using traditional breeding or cross breeding methods to acquire desired traits in plants and animals. However, advances in RDT enabled genetic changes in a more specific way and in a shorter time. This started the era of genetic manipulation and engineering of animal, plant, or microbe. Such manipulations had first started in microorganisms because of its simple organization. Genetically engineered microorganisms (GEMs) that are prepared with modern biotechnology are environment-friendly and cost-effective. For example, GEM-based production of steviol glycosides and vanillin is more beneficial as against traditional plant based methods like stevia and vanilla extracts. Several enzymes like trypsin and chymosin can also be produced using microbes. Another example is Riboflavin manufacturing is almost 100% GEM based. Nowadays food ingredients, supplements, and essential elements along with artificial sweeteners, bio enzymes and many more are produced using GEMs [21]. Success of genetic modifications was not limited to microorganisms. Plants and animals are also modified for various purposes. Antibiotic resistant tobacco and petunias varieties were developed by combining an antibiotic resistant gene and the T1 plasmid from Agrobacterium and then transforming this plasmid into the tobacco plant. France and the USA were the first countries to carry out field trials of genetically modified herbicide resistant tobacco plants. Introduction of gene for insecticidal proteins from *Bacillus thuringiensis* (Bt) into tobacco was done for the first time by Plant Genetic Systems in 1987. This was the genetically engineered insect resistant tobacco, the Bt variety [22]. Few more examples are Bt corn/maize, Bt cotton, Bt potatoes, canola with modified oil composition, resistant varieties of soybeans, squash, and tomatoes with prolonged shelf life.

The pharmaceutical industry is another sector where GMOs were used for several purposes. In 1989, the first antibody was produced using GMO. However, only a few of the genetically modified animals were approved at regulatory level. In 2003, the FDA decided that a novel genetically engineered fish (GloFish) can be marketed as a pet and not as food. In 2015 a genetically engineered salmon called the AquAdvantage salmon received approval for human consumption exclusively in the USA and Canada. This is the only genetically modified animal approved for human consumption till date [23]. But the use of genetically modified animals was more or less restricted to the laboratory to be used as model organisms. Farm animals are also modified for their improvement in production of meat, milk, wool, etc. along with getting healthy and disease resistant progeny. Biopharmaceuticals are also produced using pharming of animal modification.

Box 1.1
Dolly the sheep, considered a miracle animal, was the first cloned mammal ever to be created in a laboratory by a group of scientists at Roslin Institute, University of Edinburgh. A female sheep, Dolly was born on 5th July, 1996. Dolly was created by fusing the udder cells of a sheep with enucleated egg cells. The fusion resulted in transfer of genetic material from the udder cells into the egg cell. Following this, the egg cell was allowed to develop into an embryo and transferred to the womb of a sheep from which Dolly was born. During her lifetime, Dolly gave birth to six lambs. She led a healthy life till February 2003, when she was diagnosed with tumors on her chest. Dolly died on 14th February, 2003. The birth of Dolly started a debate regarding the ethics involved in cloning, which continues to this date.

1.6 Advances in Recombinant DNA Technology

Advances in RDT had a profound impact on society and have led to development of novel techniques and innovative products. Genetic modifications in higher eukaryotic cells were a challenge due to several limitations caused by complexity of genome organization, large gene size, etc. These obstacles encouraged the scientists to develop alternative approaches for gene modification. Comprehensive efforts and enormous experimentation helped in development of genome editing techniques. Some of the important concepts and techniques are briefly discussed below.

1.6.1 Meganucleases

Meganucleases (MegNs) are naturally occurring endo deoxyribonucleases found in prokaryotic cells as well as in mitochondria and chloroplasts allowing lateral "homing" of genes. Several homing endonucleases have been used as a tool that cleave

DNA sequences at 14–40 bp specific recognition sites other than their original wild-type targets. Very long recognition sites make these enzymes highly specific, 3′ overhangs produced after the cleavages increase frequency of joining. They have potential in development of therapeutics [24].

1.6.2 Zinc Finger Nucleases

Zinc finger nucleases (ZFNs) a gene editing tool with its zinc finger domain recognizes 3–4 bases on DNA and with its endonuclease domain it cleaves DNA and introduces double strand breaks (DSB). These proteins showed considerable application potential in cultured cells and living organisms as well [25].

1.6.3 Transcription Activator-Like Effector Nucleases

The Transcription activator-like effector nucleases (TALENs) is an artificial chimera for transcription factors and FokI restriction endonuclease. TF in the DNA-binding domain bears a repeat variable di-residue (RVD) which is responsible for specificity which recognizes one of the four nucleotides, while FokI snips the DNA. TALENs have potential use in therapeutics because of its cost-effective, safe, efficient, and target-specific nature [26]. The success rate of TALENs in genome editing is very high in livestock embryos and in plant genome engineering.

1.6.4 Clustered Regularly Interspaced Short Palindromic Repeats (CRISPR)/CRISPR-Associated Protein 9 (Cas9)

This system is a commonly used platform in the field of genome editing. This is the third-generation genomic editing tool developed in 2013 [27]. In its native form CRISPR gives immunity to the organisms by degrading foreign genetic material. CRISPR/Cas system is adapted in research laboratories for editing eukaryotic genome because of its recognition and editing capacity. A guide RNA is generated by joining a small piece of RNA and a short guide sequence which can recognize the specific target sequence in DNA. This guide RNA is then combined to CAS 9 enzyme which cuts the DNA at the targeted location. During repair the desired sequence can be added, deleted, or replaced. CRISPR genome editing is thought to be faster, less expensive, and safer. The FDA allowed CRISPR trial for hereditary childhood blindness in December 2018.

1.6.5 RNA Interference

RNA interference (RNAi), is one of the post transcriptional gene silencing techniques which depend on the breakdown of mRNA. Nowadays RNAi is developed as an efficient tool to understand mechanisms of gene regulation, gene targeting or gene therapy. The first clinical application of RNAi was to cure blindness caused by age-related macular degeneration (AMD) [28]. RNAi based therapies against viral infections, neurodegenerative diseases, and cancer are also under development. Safety and potency are the major limitations in the development of RNAi based therapeutics. The first RNAi based drug Patisiran was FDA approved in 2018 [29].

1.7 Synthetic Biology

The concept of synthetic biology was initiated by engineers who joined engineering principles with microbial cell factories. Synthetic biology utilizes gene sequences, regulator regions, gene products and living cells as standard parts to create novel gene circuits [30]. It was realized that synthesis of long DNA sequences could be a game changer in future. The use of novel chemical methods to synthesize nucleic acids evolved to BioBricks. BioBricks are arranged and assembled in multiple ways with an aim to obtain new products and cells which can be switched on or off as per our will. These cells now can show new biological characters and can carry out new functions.

1.8 Recombinant DNA Technology: An Industrial Aspect

RDT is an innovative, interdisciplinary field that influences a variety of sectors like agriculture, veterinary, medicine, pharmaceutical, fine chemicals production, environment protection products, etc. It is the leading technology which translates into solving many critical problems of society. Generation of new therapeutics, engineering new crops and livestock, production of transportation fuels and chemicals for industrial manufacturing was possible with the advancement of RDT. Modern biotechnology can be recognized as the second revolution in the medical and industrial sector as it has increased our ability to obtain pure bio-based products on a large scale in a cost-effective manner. Many biotechnology companies, like Genentech, started to commercialize biotechnology in the last decade. The global biotechnology market value is expected to expand at a compound annual growth rate (CAGR) of 15.83% from 2021 to 2028. Many sectors are going to contribute to the growth. As expected, the health application segment held the largest share of 48.64% in 2020 and this trend is expected to continue. One of the significant roles will be played by innovative biopharmaceuticals companies as around 40% drugs are biotech derived. Agricultural companies are also contributing to market growth by bringing agricultural innovations for improving productivity with sustainable solutions.

All these innovations are in order to increase crop productivity, get stress tolerant plants, allow better diagnostics, make available affordable drugs, and protect us from epidemics such as the very recent COVID-19. These multidisciplinary applications and growth rate in the market at global level highlights the bright future of RDT [31].

1.9 Summary

rDNA techniques marked a paradigm shift in the world of life sciences. It has altered the approach of a researcher to a problem. Construction of new rDNA and its introduction into microbes, plants, and experimental animals is a standard practice in research laboratories nowadays. The technology has allowed us to treat challenging genetic disorders, boost agriculture productivity, biofortified the crops, brought humongous change in forensic science, bioremediation, etc. However, the ethical dilemma over the increase in commercialization of rDNA organisms has always been a concern and subject of debate. But accumulation of research and experience allow us to understand the potential risks and predict the outcomes associated with rDNA organisms, thereby helping us to take appropriate safety measures for the betterment of mankind.

Self Assesment

Q1. What are the major discoveries that led to the development of RDT?

A1. Unraveling the concepts of genetics, identifying DNA as the hereditary factor, isolation of plasmids, discovery of restriction enzymes, deciphering the genetic code, etc. are some of the important discoveries that paved the way for development of RDT.

Q2. Describe the first recombinant protein approved by FDA

A2. Humulin is recombinant insulin produced in 1979. Eli Lilly and Company started a joint-venture with Genentech for the development and production of recombinant human insulin which was named Humulin. Humulin was approved by FDA in 1982 and became the first biotechnology product launched in the market.

Q3. Explain any two types of nucleases relevant to RDT.

A3. Zinc finger nucleases are artificially engineered restriction enzymes for genome editing. Zinc finger nuclease is a hybrid heterodimeric protein where each subunit contains several zinc finger domains and a Fok1 endonuclease. With its zinc finger domain, it recognizes 3–4 bases on DNA and with its endonuclease domain it cleaves DNA and introduces double strand breaks.

The TALEN is an artificial chimeric protein made by joining a nonspecific FokI restriction endonuclease domain to a DNA-binding domain recognizing an arbitrary base sequence. This DNA-binding domain comprises highly conserved repeats derived from transcription activator-like effectors (TALE). In the DNA-binding domain there is a repeat variable di-residue (RVD) which is responsible for specificity.

References

1. Bhatia S, Goli D. Introduction to pharmaceutical biotechnology. In: Basic techniques and concepts, vol. 1. Bristol: IOP Publishing; 2018.
2. Primrose B, Twyman R. Principles of gene manipulation and genomics. Wiley; 2013.
3. Cohen SN, Chang ACY, Boyer W, Helling B. Construction of biologically functional bacterial plasmids in vitro. Proc Natl Acad Sci U S A. 1973;70(11):3240–4.
4. Mendel G. Versuche uber Pflanzen-Hybriden. Verh. Naturforsch. Ver. Brunn 4 3-47 English translation in 1901. JR Hortic Soc. 1886, 26:1–32.
5. Sutton S. The chromosomes in heredity. Biol Bull. 1903;4(5):231–50.
6. Morgan H. Sex limited inheritance in Drosophila. Science. 1910;32(812):120–2.
7. Morgan H. The theory of the gene. New Haven: Yale University Press; 1926.
8. Garrod A. The incidence of alkaptonuria: a study in chemical individuality. Lancet. 1902;160(4137):1616–20.
9. Beadle W, Tatum L. Genetic control of biochemical reactions in Neurospora. Proc Natl Acad Sci U S A. 1941;27(11):499.
10. Lewis B. Pseudoallelism and gene evolution. In: Cold Spring Harbor symposia on quantitative biology, vol. 16. Cold Spring Harbor Laboratory Press; 1951. p. 159–74.
11. Benzer S. The elementary units of heredity. In: McElroy WD, Glass B, editors. The chemical basis of heredity. Baltimore: The Johns Hopkins University Press; 1957. p. 70–93.
12. Gamow G. Possible relation between deoxyribonucleic acid and protein structures. Nature. 1954;173(4398):318.
13. Yanofsky C, Crawford P. The effects of deletions, point mutations, reversions and suppressor mutations on the two components of the tryptophan synthetase of Escherichia coli. Proc Natl Acad Sci U S A. 1959;45(7):1016.
14. Lobban E, Kaiser D. Enzymatic end-to-end joining of DNA molecules. J Mol Biol. 1973;78(3):453–71.
15. Mertz E, Davis W. Cleavage of DNA by R1 restriction endonuclease generates cohesive ends. Proc Natl Acad Sci U S A. 1972;69(11):3370–4.
16. Jackson A, Symons H, Berg P. Biochemical method for inserting new genetic information into DNA of Simian Virus 40: circular SV40 DNA molecules containing lambda phage genes and the galactose operon of Escherichia coli. Proc Natl Acad Sci U S A. 1972;69(10):2904–9.
17. Chang C, Cohen N. Genome construction between bacterial species in vitro: replication and expression of Staphylococcus plasmid genes in Escherichia coli. Proc Natl Acad Sci U S A. 1974;71(4):1030–4.
18. Morrow F, Cohen N, Chang C, Boyer W, Goodman M, Helling B. Replication and transcription of eukaryotic DNA in Esherichia coli. Proc Natl Acad Sci U S A. 1974;71(5):1743–7.
19. Saltepe B, Kehribar S, Su Yirmibeşoğlu S, Şafak Şeker O. Cellular biosensors with engineered genetic circuits. ACS Sens. 2018;3(1):13–26.
20. Sandeep V, Parveen J, Chauhan P. Biobetters: the better biologics and their regulatory overview. Int J Drug Regul Aff. 2016;4(1):13–20.

21. Hanlon P, Sewalt V. GEMs: genetically engineered microorganisms and the regulatory oversight of their uses in modern food production. Crit Rev Food Sci Nutr. 2020;61(6):959–70.
22. Vaeck M, Reynaerts A, Höfte H, Jansens S, De Beuckeleer M, Dean C, Leemans J. Insect resistance in transgenic plants expressing modified Bacillus thuringiensis toxin genes. Nature. 1987;328:33–7.
23. Cotter J, Perls D: "Genetically Engineered Animals: From Lab to Factory Farm", FRIENDS OF THE EARTH, September 2019 (2019-09-01), pages 1–41, XP055722286 (TPO)
24. Petersen B, Niemann H. Molecular scissors and their application in genetically modified farm animals. Transgenic Res. 2015;24(3):381–96.
25. Carroll D. Genome engineering with zinc-finger nucleases. Genetics. 2011;188(4):773–82.
26. Khan SH. Genome-editing technologies: concept, pros, and cons of various genome-editing techniques and bioethical concerns for clinical application. Mol Ther Nucl Acids. 2019;16:326–34.
27. Cong L, Ran A, Cox D, Lin S, Barretto R, Habib N, Zhang F. Multiplex genome engineering using CRISPR/Cas systems. Science. 2013;339(6121):819–23.
28. Tolentino M. Interference RNA technology in the treatment of CNV. Ophthalmol Clin N Am. 2006;19(3):393–9.
29. Setten L, Rossi J, Han P. The current state and future directions of RNAi-based therapeutics. Nat Rev Drug Discov. 2019;18(6):421–46.
30. Xue S, Yin J, Shao J, Yu Y, Yang L, Wang Y, Ye H. A synthetic-biology-inspired therapeutic strategy for targeting and treating hepatogenous diabetes. Mol Ther. 2017;25(2):443–55.
31. Hossain G, Saini M, Miyake R, Ling H, Chang M. Genetic biosensor design for natural product biosynthesis in microorganisms. Trends Biotechnol. 2020;38(7):797–810.

Chapter 2
Vectors: Guide to Gene Delivery Vehicles

Shalmali Bivalkar, Rajeev Mehla, Aruna Sivaram, and Nayana Patil

2.1 Introduction

A "vector" in Latin means a carrier. The word vector was first used by astronomers, way back in the eighteenth century to explain the revolutions of planets around the Sun. While in Geometry, a vector is an object with a magnitude and a direction. A vector is something that allows transfer of material from one point to another in a specific direction. On the similar lines, a cloning vector allows transfer of genetic material from one organism (cell) to another. In the last four decades recombinant DNA technology (RDT) has revolutionized various fields ranging from agriculture, food industry to health sector. To present a very relevant example, where development of an attenuated vaccine by classical approach took anywhere from 10 to 15 years in the past, RDT has made it possible to develop a recombinant COVID-19 vaccine in a matter of months in response to a raging pandemic in 2020. Similarly, recombinant insulin, gene therapy for correction of congenital diseases, pest-resistant crops like *Bt* cotton are all products of RDT.

RDT deals with transfer of DNA from one cell to another or from one species to another conferring novel characteristics to the recipient. Isolation of the DNA fragment of interest is the starting step of RDT. The DNA fragment can be the coding

S. Bivalkar
National Centre for Cell Science, Pune, Maharashtra, India

R. Mehla
Serum Institute of India, Pune, Maharashtra, India
e-mail: rajeev.mehla@seruminstitute.com

A. Sivaram · N. Patil (✉)
School of Bioengineering Sciences & Research, MIT ADT University,
Pune, Maharashtra, India
e-mail: aruna.sivaram@mituniversity.edu.in; nayana.patil@mituniversity.edu.in

N. Patil, A. Sivaram, *A Complete Guide to Gene Cloning: From Basic to Advanced*,
Techniques in Life Science and Biomedicine for the Non-Expert,
https://doi.org/10.1007/978-3-030-96851-9_2

sequence for a protein or RNA (transgene) or it could be a synthetic piece of DNA. Once the foreign DNA fragment is isolated it is amplified by conventional polymerase chain reaction (PCR). If such a DNA fragment is introduced into the host cell, it will be degraded by the cell's defense mechanism identifying it as non-self. Hence, to safely deliver the DNA fragment, a vehicle that will protect the DNA fragment, deliver it and make it functional in the host cell is necessary. These vehicles are the cloning vectors, first such vectors were developed from plasmids.

Plasmid DNA molecules were first discovered in bacteria and were later modified for use as cloning vectors. Cloning vector acts as a vehicle to transfer the desired piece of DNA into the target cells. Like plasmid, components of bacteriophage genomes, yeast chromosomes and viral genomes were later studied and used in RDT for various applications. Cloning vectors are used for multiple purposes including transfer and expression of transgene into a host cell for recombinant protein production, conferring special characteristics to the host cell (e.g., metabolic, structural, or survival benefit), controlling expression of cellular genes, genome editing, etc. In some instances, cloning vectors allow preservation of DNA fragments that are inserted into the vector allowing library preparation for the cellular genomes. A cloning vector must have following characteristics:

- Ability to replicate in a specific host cell (e.g., bacteria, yeast, plant, mammalian cell, etc.)
- Contains a genetic marker for selection of cells carrying plasmid.
- Unique restriction enzyme sites to allow insertion of foreign DNA.
- Minimum amount of nonessential DNA sequences.

Cloning vectors vary in structure, design, size, their insert carrying capacity and replication efficiencies. Different types of cloning vectors including plasmids,

Table 2.1 Types of cloning vectors

Sr. no.	Type of vector	Origin	Applications
1.	Plasmid	Bacterial plasmid	DNA insert amplification, storage, transgene expression
2.	Cosmid	Bacteriophage lambda and plasmid	DNA sequencing, protein expression, cDNA library
3.	Phagemid	Bacteriophage and plasmid	Mutagenesis, sequencing, single-stranded DNA synthesis (probes, DNA standards)
4.	YAC	Yeast chromosomes	Genomic library construction, generation of transgenic cell lines and animals
5.	BAC	Bacterial F plasmid	Genomic library preparation, genome mapping
6.	HAC	Human chromosome	Transgenic animals, gene therapy, regulation of gene expression
7.	Lentiviral vector	Lentivirus	Transgenic animals, gene therapy, regulation of gene expression
8.	Adenoviral vector	Adenovirus	Gene therapy, viral vaccines

cosmids, phagemids YACs and BACS are available (Table 2.1). Each of these vectors have their own pros and cons and are used for specific applications based on their properties.

We will now look at each type of cloning vector in finer detail. Clear understanding about cloning vectors would help us in designing better and efficient cloning strategies and choosing the right vector for a specific experiment.

2.2 Plasmids

In 1952, a young scientist named Joshua Lederberg was working on cytoplasmic heredity in bacteria when he used the term "plasmid" for the first time [1]. He would not have known at that point of time that plasmid would eventually become the workhorse of RDT. Later, plasmids were found in other prokaryotic species including bacteria and archaebacteria. Plasmids are extrachromosomal circular, self-replicating, double-stranded DNA molecules present in bacteria. At the time of cell division, plasmid DNA along with the genomic DNA is segregated into daughter cells ensuring its passage to the next generation daughter cells [2]. Some plasmids carry code for their own transfer from one cell to another (F plasmid), while others code for proteins for antibiotic resistance (R plasmids) or for metabolism of unusual chemicals (degradative plasmids). Plasmids with sequences that are still unexplored or do not constitute for any known functional gene are termed cryptic plasmids. Plasmids vary in size from nearly 1 kilobase (kb) to over 500 kb. R plasmids were first DNA vectors shown to transfer genetic material from one bacterial cell to

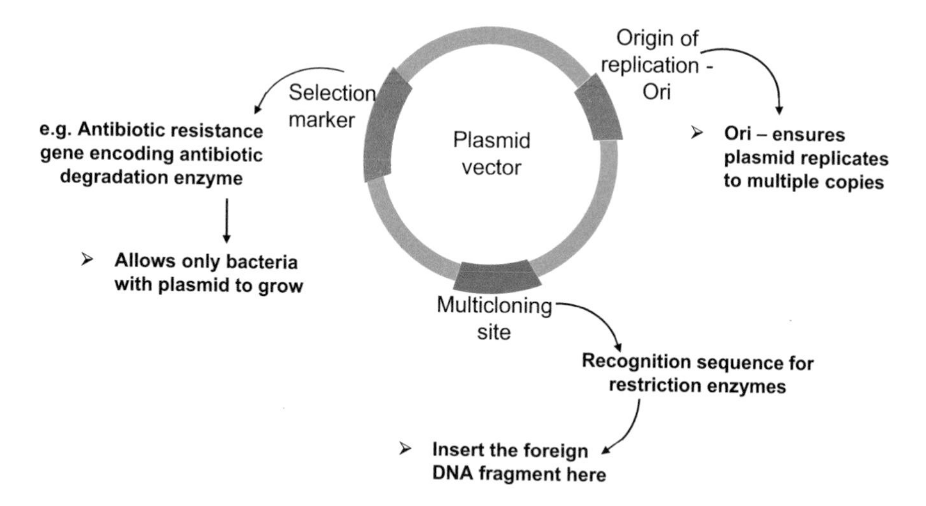

Fig. 2.1 Components of a basic plasmid vector and their function. Basic components of a plasmid include origin of replication (Ori), selection marker gene promoting survival of plasmid bearing host cells, and multi-cloning site composed of restriction enzyme recognition sites where foreign DNA fragment is inserted

Table 2.2 Basic components of a plasmid vector

Sr. no.	Component	Function
1	Origin of replication (Ori)	Responsible for assembling replication machinery for plasmid replication independent genomic DNA replication
2	Selection marker	Required for selection of host cells that contain plasmid vectors. Most of the time it is an antibiotic resistance gene
3	Multiple cloning site (MCS) or polylinker	Recognition sites for restriction enzymes where DNA fragment is inserted while cloning in plasmid

another and confer the associated antibiotic resistance phenotype onto the recipient bacteria [3]. A basic design of plasmid vectors is depicted in Fig. 2.1.

Since their discovery plasmids have proved to be a versatile tool in basic and applied science, biotechnological advances from production of therapeutic, recombinant proteins to development of pest-resistant crops. Naturally occurring plasmids are altered to make them suitable for application in RDT. Their size is reduced by removing unwanted sequence elements, and certain sequences are added to enhance their replication or improve the expression of transgene. Plasmids can carry a DNA insert of up to 10 kb. Table 2.2 summarizes basic components of a plasmid vector.

Origin of Replication (Ori) and Plasmid DNA Replication

An origin of replication consists of characteristic DNA sequences recognized by DNA replication machinery of respective bacterial species. Ori provides a site for the assembly of different components of host replication machinery. Plasmids can replicate on their own independent of the host cell genome and follow a rolling circle mechanism of replication due to their circular morphology [4]. One of the plasmid strands is nicked and the intact strand is first replicated displacing the nicked strand, following which the nicked strand is also replicated to form two double-stranded circular plasmid molecules. Plasmids designed for cloning are equipped with Ori sites that allow efficient recruitment of the host replication machinery and thus distribution into the daughter cells. Certain Ori sequences are specific to bacterial species and are recognized by only their replication machinery components (narrow host range plasmids), while some Ori sequences are functional in multiple bacterial species (broad host range plasmids). Often more than one Ori is included in plasmids to permit their use in more than one host cell. The copy number of the plasmid can be carefully regulated based on the choice of Ori sequence. A plasmid vector is categorized as high or low copy number based on its replication efficiency which is a function of regulatory mechanisms working on the Ori in each host system.

Selection Marker Gene

Selection marker is a gene carried by the plasmid that confers selective advantage to the plasmid bearing host cells e.g., an antibiotic resistance gene or an enzyme coding gene that grants metabolic advantage. In naturally occurring plasmids this gene renders survival benefit to the cells, thus favoring their population growth in the relevant environment (e.g., multi-drug resistant *E. coli*). This feature is used in RDT to positively select the cells that contain the plasmid over the cells that do not possess it. Based on the method used for transfer of plasmid DNA into the host cells either bacterial (transformation) or eukaryotic (transfection), a certain proportion of cells receive the plasmid and are termed transformants. Selection marker gene expression plays a role in specifically selecting transformants. For example, in the case of antibiotic resistant genes, transformants can replicate in presence of the respective antibiotic and are selected, whereas cells without plasmid are eliminated by the action of antibiotic added to the culture medium.

Commonly used selection marker genes in plasmids include ampicillin resistance gene (*amp*^{R}) expressing beta-lactamase enzyme, kanamycin resistance gene (*kan*^{R}), tetracycline resistance gene (*tet*^{R}), etc. Sometimes plasmids may also carry fluorescent protein expressing genes acting as a selection marker for transformants by visual methods e.g., Green Fluorescent Protein (GFP) or Red Fluorescent Protein (RFP).

Multiple Cloning Site

Multiple Cloning Site (MCS), also called polylinker, is a region in the plasmid consisting of a series of recognition sequences for multiple restriction enzymes. Restriction enzymes can cleave double-stranded DNA at specific sequence elements. Plasmid DNA can be cut at specific restriction site allowing insertion of foreign DNA fragments (also called insert) through ligation. Restriction enzymes are chosen strategically to create blunt or sticky (cohesive) ends on both plasmid and exogenous DNA insert. For insertion of foreign DNA fragment into the cloning vector, vector and the fragment are both cut or digested with the same restriction enzyme such that the ends of the vector and the foreign DNA insert are compatible with each other and can be ligated to form a recombinant vector with foreign DNA inserted in it.

To understand clearly about the components of the plasmid vector, let us look at a classic example of a plasmid vector, named pBR322. Plasmid pBR322 carries Ori derived from *E. coli* plasmid ColE1 that ensures a high copy number of 15–20 copies per cell. The plasmid has two antibiotic resistance genes as selection markers. First, ampicillin resistance gene (*amp*^{R}) is derived from the Tn3 transposon and second, tetracycline resistance gene (*tet*^{R}) is derived from pSC101 plasmid. MCS is included in the vector to allow easy insertion of foreign DNA fragments.

The selection markers possessed by pBR322 work in two ways:

- Bacterial cells carrying plasmid expressing *amp*^{R} get selected when grown in a culture medium containing ampicillin (positive selection). However, bacteria with empty plasmid vectors without the insert (empty vector) will also be able to grow. Thus, selection marker *amp*^{R} is useful to differentiate between transformants and non-transformants.
- To ensure selection of bacteria with plasmid containing the inserted DNA fragment (recombinant plasmid), the second selection criterion is employed. Tetracycline resistance gene (*tet*^{R}) contains recognition sites for restriction enzymes BamHI and HindIII. When the foreign DNA fragment is inserted at these restriction sites, the *tet*^{R} gene is inactivated (insertional inactivation).

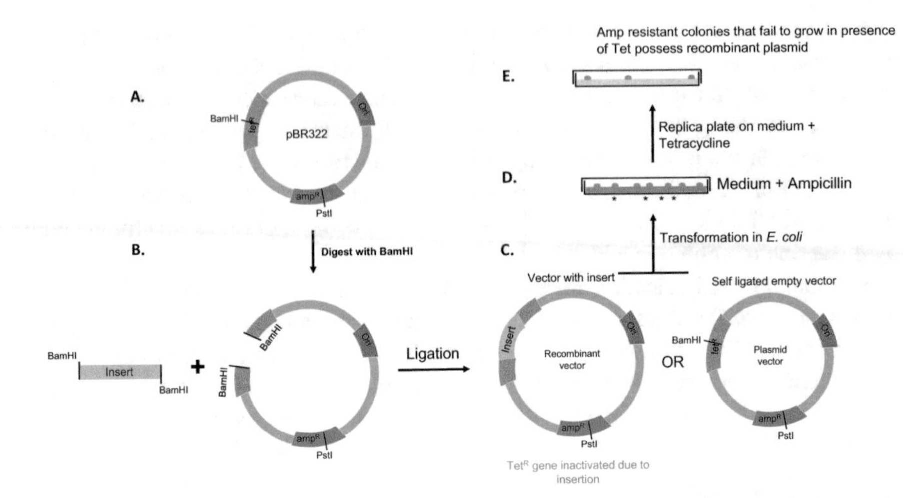

Fig. 2.2 Cloning in pBR322 Vector. pRB322 is a plasmid vector with a ColE1 Ori, two selection marker gene that work through positive and negative selection of transformants. A basic cloning strategy and mechanism of selection of transformants is depicted in the figure. When the foreign DNA fragment is inserted BamHI restriction site tetR gene is inactivated due to insertion of DNA fragment. Following transformation of E. coli with ligation product, in the first step selection of transformants is based on ampicillin resistance. All the bacteria with plasmid in them can grow in presence of ampicillin due to intact ampR gene. In the second step of selection, tetracycline added medium is used for replica plating the colonies grown on ampicillin medium. Only the bacteria with tetR gene intact can now grow thus eliminating the colonies with recombinant plasmid. Thus the clones which lost tetracycline resistance but retained ampicillin resistance are the ones containing plasmid with the foreign DNA fragment

Hence, when the transformed bacteria are first grown in ampicillin containing medium, all of the cells with plasmid can grow. When these colonies are replica-plated on the medium containing tetracycline, only the bacterial cells with plasmids lacking insertion and an intact tetracycline resistance gene can grow (Fig. 2.2). Similar selection mechanism of antibiotic sensitivity due to insertional inactivation can be used when a DNA fragment is inserted in the restriction sites present in the ampicillin resistance gene.

pBR322 vector has been used to construct better cloning vectors. pUC series of vectors was developed in 1982 and had three additional characteristics as compared to pBR322 [5] that offered many advantages:

- The two-step screening method used for selection of recombinant clones of pBR322 was time consuming and error prone. pUC vectors were equipped with a novel single step screening method for transformants—blue-white screening. pUC vector contains a gene that encodes an amino-terminal portion of *E. coli* beta-galactosidase enzyme called alpha peptide. Bacterial host cells used in this case express a nonfunctional form of beta-galactosidase enzyme, the omega peptide. The omega peptide becomes functional when put together with alpha peptide and the process is termed as alpha complementation. The feature is exploited by the scientists for construction of insertional inactivation-based selection markers. A multi-cloning site was inserted in the coding sequence of alpha peptide such that insertion of DNA fragment within this MCS leads to insertional inactivation of alpha peptide gene. A nonfunctional peptide is translated from such a gene leading to loss of alpha complementation with omega peptide. When the transformed bacteria are supplied with a chromogenic (color producing) substrate for beta-galactosidase, the cells without insert and thus with alpha complemented beta-galactosidase can catalyze the substrate producing blue colonies. The bacterial cells with a recombinant plasmid containing inserts within the alpha peptide gene fail to catalyze the substrate and form white colonies.

- Presence of a single mutation (G to A) within the Ori sequence resulted in the elimination of negative regulation on the plasmid replication. The mutation resulted in an improved replication efficiency from around 20 copies/cell in case of pBR322 to around 500 copies/cell.
- pUC vectors contained a multiple cloning site or a polylinker. A synthetic piece of DNA containing recognition sites for multiple restriction enzymes placed next to each other was created. This polylinker was inserted within the coding sequence of the alpha peptide gene in such a way that it did not alter the expression or function of the expressed alpha peptide. However, insertion of foreign DNA fragments into any of these restriction sites would disrupt the alpha peptide gene leading to insertional inactivation.

Additional important components of a plasmid may include promoter region required for expression of the transgene, sequence elements enhancing expression of inserted transgene e.g., enhancers, exons, or other untranslated region portions derived from naturally occurring genes. Similarly, in order to facilitate DNA sequencing of the insert carried by the vector, a primer binding site is included in the cloning vectors. Standard universal primers (with unique sequences) can be used to sequence the region with foreign DNA insert to confirm cloning. The field of RDT is constantly growing and new functional elements are being added to make plasmids more applicable. Plasmids are efficient and robust vehicles for carrying DNA sequences from one host to the other. Today, due to novel protocols and availability of tools, use of plasmids, their manipulation, amplification, and cloning has become much easier.

One of the first widely used plasmids, isolated from *E. coli* and still very popular in RDT is pBR322 plasmid. This vector was created in the laboratory of Herbert Boyer at University of California, San Francisco. In pBR322, p stands for plasmid, B is from Bolivar and R from Rodriguez. The vector was named after its creators Francisco Bolivar Zapata, the postdoctoral researcher, and his coworker Raymond L. Rodriguez. The number 322 represents the clone number. pBR322 is around 4.3 kb (4300 base pairs) in length and is constructed from components derived from various naturally occurring plasmids and DNA fragments (Fig. 2.2).

Similarly, pUC series of vectors were developed by Messing and coworkers using some components from pBR322. In pUC, p stands for plasmid while UC stands for University of California. pUC18 is one of the plasmids in the pUC series of vectors. Thus, plasmids were named in a specific way which has evolved over the years.

2.2.1 Lambda Bacteriophage-Based Vectors and Cosmids

Plasmids, as discussed in the previous section, provide a promising vehicle for DNA cloning however, initially they could only be used for carrying smaller inserts. Moreover, higher sized inserts led to poor transformation efficiency. Scientists were looking for alternative tools to overcome these limitations. The breakthrough came in the 1970s with the use of bacteriophages as vectors. Bacteriophages could transfer their own genome into bacterial host cells. Similar to the plasmid transformation method in bacteria, bacteriophages were used for introduction of large DNA fragments into the bacterial cells.

Bacteriophages (also called phages) are viruses that propagate themselves by exploiting bacterial host cell machinery. During their life cycle, phage proteins are translated which assemble into phage head and tail. Phage DNA undergoes multiple cycles of replications to produce numerous copies of the phage genome. When the phage proteins and genome copies are present in ample amounts within the host cell, the genome is packaged into the assembled phage heads and new phages are released from the host cell causing its lysis (lytic cycle). In certain situations, the phage genome is retained in the bacterial host cell without its replication or protein expression. Phage surviving in such a dormant state within the host cell is termed lysogenic infection. Upon induction by specific bacteriophage proteins, lytic infection is initiated and new phages are released form the host cells. Lytic or lysogenic life cycle can be determined from the plaques formed in the bacterial lawn culture. Lytic life cycle results in clear plaques due to lysis of bacterial cells whereas lysogenic life cycle results in turbid plaques.

One such well-studied bacteriophage is lambda (λ) phage. Esther Lederberg in the 1950s discovered and studied λ phage which infects *E. coli*. Its genome is synthesized as a concatemer (polymer of identical repeating genome units) with multiple copies of phage DNA separated by cohesive end (COS) sites. It was determined that DNA is found in polymeric form (concatemers) with monomers arranged head to tail separated by cohesive end sites (COS sites). COS sites are composed of a specific sequence of 22 bp within which a recognition sequence for λ phage terminase enzyme is located [6]. Terminase is an endonuclease that interacts with the COS sites and makes a double-stranded cut within COS sites, resulting in the DNA ends with single-stranded overhangs. The overhang on one side of the genome is complementary to the overhang created on the other end of the genome. Such λ genome of unit length decorated by COS sites at ends is packaged into the phage heads. Upon infection of *E. coli* cells, a linear λ genome is introduced into the cytoplasm of the host cell followed by circularization of the genome guided by the complementary base pairing of the COS sites on the ends. Researchers discovered that replacement of certain portions of the phage genome with foreign DNA while retaining intact COS sites, did not hamper packaging of the genome into the phage head. This knowledge led to the development of λ phage-based vectors and cosmids [7].

Lambda (λ) Bacteriophage-Based Vectors

λ vectors were developed by removal of genes unnecessary for completion of lytic phage life cycle, the space thus created was available for insertion of foreign DNA. To make the vector more suitable for cloning unwanted restriction sites were removed without affecting the phage protein coding sequences and new restriction sites were added.

Lambda vectors can be of two types based on their design (Fig. 2.3):

- Insertion vector: Foreign DNA is inserted in the phage genome at a single target site specifically cut by a restriction enzyme. The size of the λ insertion vector is about 20% smaller than the wild type phages as a result of removal of many genes unnecessary for the lytic life cycle of the phage. Insertion of a foreign DNA fragment of appropriate size brings back the genome size into the range required for proper packaging into the phage heads [8]. Lambda insertion vectors can accommodate from 5 kb to 11 kb, depending on the design of the vector. Examples of λ insertion vector is Lambda gt10 which can carry upto 8 kb of foreign DNA insert at a unique EcoRI restriction enzyme site located within the *cI* gene of λ phage. Recombinant phages show insertional inactivation of *cI* gene, resulting in clear plaques rather than turbid plaques; 2. Lambda ZAPII vector with an insert capacity of 10 kb has a polylinker situated within the *lac z'* gene (coding for beta-galactosidase enzyme) inserted in the phage genome. Insertion of DNA fragments within *lac z'* gene results in blue plaques rather than clear plaques.

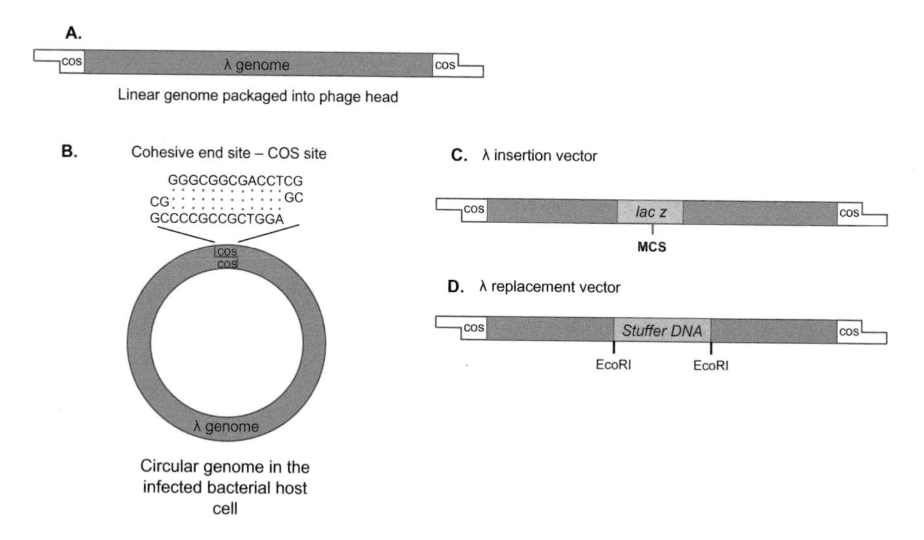

Fig. 2.3 λ phage-based vectors. (**a**) Bacteriophage lambda genome in its linear form can be packaged into phage head. (**b**) The circular form of the lambda genome can be seen in bacterial hosts. Complementary sequences in the COS sites are responsible for circularization of the lambda genome within the bacterial cells. (**c**) Two types of lambda cloning vectors—λ insertion vector where MCS is provided within a lacZ gene added to the phage genome and (**d**) λ replacement vector where some portion of the phage genome is replaced by a stuffer DNA which can be removed to insert the foreign DNA fragment in its place during cloning

- Replacement vector: A piece of DNA (Stuffer fragment) in the vector is replaced with foreign DNA. Replacement vectors possess two restriction enzyme target sites flanking a fragment of nonessential DNA called stuffer. When a linear replacement vector is digested by appropriate restriction enzymes, it removes the stuffer DNA and the left and right arms are generated. In the absence of the foreign DNA fragment, left and right arms get ligated and the size of the empty vector is much smaller than what is required for packaging into the phage head. Hence, packaging of only the recombinant vector (the one containing the DNA fragments) is ensured. For example, LambdaEMBL4 is a vector with 42 kb size. The stuffer fragment in it is 14 kb. After removal of the stuffer fragment if the two arms of the vector are ligated, they stand at a size of 28 kb which is too small to be packaged into a λ phage head. Advantage of λ replacement vector is its ability to carry larger DNA fragments as compared to λ insertion vector.

Once the recombinant λ phage vectors are created, they are packaged into the phage heads in vitro. Components of λ phage head are produced using mutant strains of *E. coli* that complement (depend on) each other for mature head formation and packaging when their lysates are mixed along with the recombinant λ vector. Packaging of a recombinant λ phage vector was a complicated, multistep protocol that required multiple bacterial strains. To overcome this difficulty Collins and coworkers in 1978 described cosmid vectors which were a combination of plasmid and λ phage components.

Cosmid

Use of phage elements for improving cloning vectors was established by the development of cosmid. Unlike lambda vectors, cosmids were created by adding only the λ COS sites to the plasmid. Cosmids do not carry any other lambda phage genes. As a result, cosmids carry foreign DNA fragments up to 45 kb between two COS sites. Presence of COS sites allows packaging of cosmids into the phage head in *E. coli* carrying a helper phage. Helper phage expresses all necessary phage genes for lytic infection but lacks packaging signals (COS sites). resulting in newly synthesized phages packaged with cosmids carrying foreign DNA fragments. Alternatively, cosmids carrying the DNA insert can be packaged in vitro using phage proteins (Fig. 2.4). These phages can be used to introduce the vector into the required host cells through infection with high efficiency. Other than COS sites, cosmids share the rest of the components with plasmids that are required for replication and selection of positive transformants (cells containing the cloning vector).

2.2.2 *M13 Bacteriophage-Based Vectors and Phagemids*

Like λ phage, other bacteriophages were also tested for their ability to function as a cloning vector. M13 is a filamentous phage that infects *E. coli*. The name "filamentous" is derived from the long, thin ribbon-like shape of the phage. It consists of a

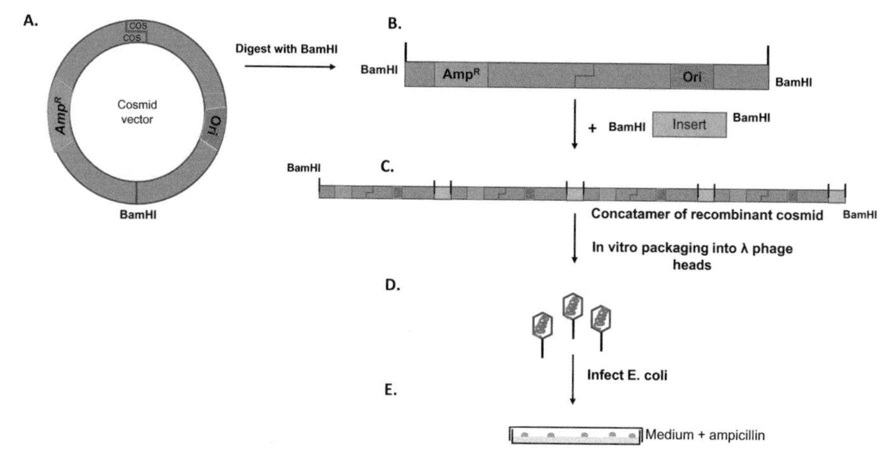

Fig. 2.4 Cosmid vector and basic steps for cloning in cosmid vector. Cosmid vector consists of a plasmid backbone and an additional COS site from λ bacteriophage. For cloning a foreign DNA fragment in a cosmid vector, first the vector is linearized using restriction enzyme BamHI. DNA fragments to be inserted into the vector are also digested with BamHI to have compatible ends with the vector. When the linearized vector and insert are ligated, they form a concatemer of the phage genome with a DNA fragment inserted into the vector. Concatemeric genome is used for in vitro packaging into the phage heads, where unit length of recombinant phage genome is packaged to form recombinant phages. These phages can be used for infection of suitable *E. coli* hosts for further screening

single-stranded circular DNA molecule as its genome and specifically infects bacterial cells with F pilus. Upon infection, phage enters the host cells and the phage genome is converted to double-stranded DNA molecules. Phage genes are expressed into proteins and the phage genome is replicated. One of the phage proteins (encoded by gene 2) forces the replication machinery to replicate only one of the two strands, the plus strand. Plus (+) strand is then packaged into the phage heads during assembly for the release of new phage particles. The host cell remains alive during this entire process and divides slowly due the metabolic burden. Following the discovery of M13 phage, in 1977, Joachim Messing a graduate student at Max-Planck Institute in Munich uncovered the potential of M13 as a cloning vector by inserting a component of *lac* operon into the intergenic region of M13 genome [9]. Multiple versions of M13 cloning vectors were then developed (Fig. 2.5) and used for various applications, mainly where a single-stranded DNA product was required.

Phagemids are plasmids that carry origin of replication from a regular plasmid and an additional origin of replication from a filamentous bacteriophage such as f1, fd, or M13. Phagemids, unlike the M13-based vectors, do not contain any phage genes. Presence of a phage Ori site allows the phagemid to undergo single strand replication mode in presence of a helper phage in the host cell (Fig. 2.6). Otherwise, the phagemid behaves the same as a plasmid when transformed into a suitable host cultured in selective condition. Having phage as the carrier of a cloning vector allows efficient introduction of vectors into the bacterial host cells through phage infection.

Box 2.1
If a certain protein or peptide is to be studied for its interactions, functions, and physicochemical properties, it is essential that the protein is available in ample amounts. To clone the sequence coding for the required protein or peptide one should know the sequence first. Phage display system allows revelation of genotype–phenotype linkage. Through phage display one can isolate the sequence coding the protein or peptide of interest. Phage display system allows isolation of specific sequence coding for the specific protein or peptide of interest. Phage display system is utilized for studying protein–protein interaction, receptor–ligand binding, antibody–antigen binding and also for improving binding affinity [10]. In the phage display technique, DNA sequences coding for peptides or the cDNA library coding for multiple peptides to be screened are cloned in M13-based vectors. The peptide-coding sequence is cloned in frame with one of the phage coat protein expressing genes, resulting in a fusion protein being displayed on the surface of the phage. Phages are packaged using suitable bacterial strain or in vitro. Recombinant phages thus produced are subjected to a screening process called panning where their interaction with the molecule of interest is tested. The phages that express the right peptide are selected during the panning process due to their affinity for the ligand or interacting protein. Rest of the phages get eliminated. Selected phages are used for extracting the DNA and sequencing to reveal the sequence coding for the protein or peptide of interest. Phage display technique has proved to be useful for clone selection in monoclonal antibody production, epitope mapping in understanding antigen–antibody interaction, receptor and ligand identification, directed evolution of proteins like enzymes and antibodies, drug discovery—identifying drugs that interact with specific proteins to block their function or interaction as a therapeutic option.

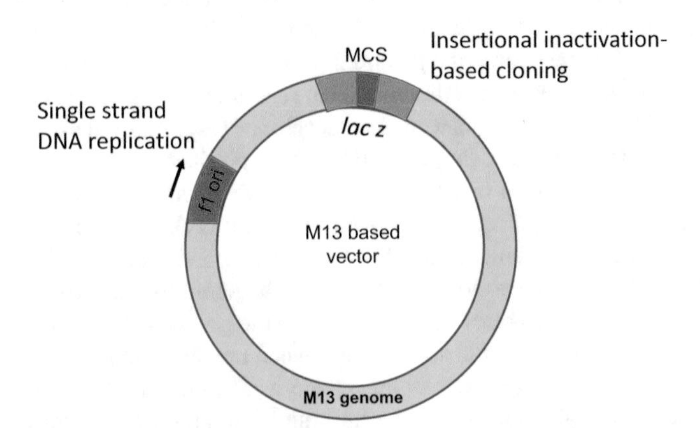

Fig. 2.5 Basic components of a M13-base cloning vector. M13-based cloning vector is equipped with phage f1 Ori site. This Ori site is responsible for single-stranded DNA replication. The selection marker is based on insertional inactivation of lac Z gene which contains the MCS

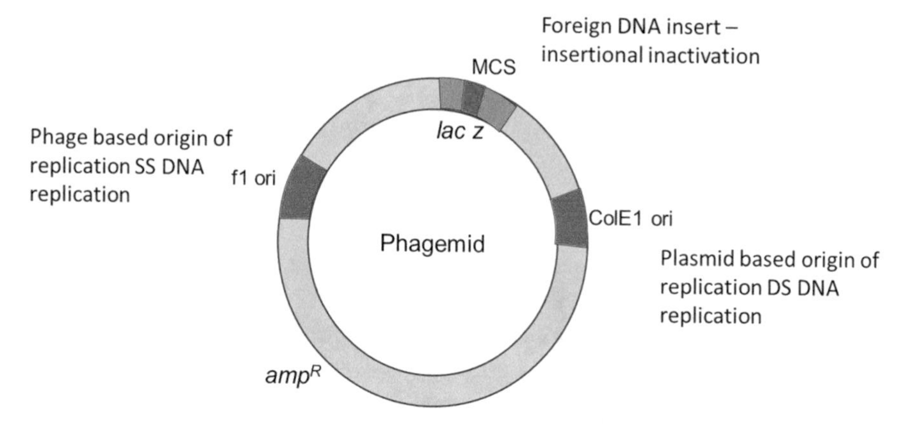

Fig. 2.6 Basic components of phagemid vector. A phagemid contains ColE1 Ori in addition to f1 Ori allowing double-stranded DNA replication and single-stranded DNA replication. It has two selection marker genes—ampR for antibiotic resistance and lacZ with an MCS within it. Insertion of DNA fragments in lacZ can be detected through insertional inactivation of the gene

2.3 Artificial Chromosomes

Artificial chromosome vectors are high-capacity cloning vectors with ability to carry large DNA fragments stably. They were termed chromosomes due to their size in case of BACs and also due to the components they share with eukaryotic chromosomes (YACs). Main purpose of artificial chromosomes was to insert large genomic fragments to construct libraries and maps of eukaryotic and mammalian chromosomes.

2.3.1 Yeast Artificial Chromosomes (YACs)

Inability to carry large DNA fragments was one of the major limitations of plasmid-based cloning vectors. Eukaryotic chromosomes are DNA structures that carry a multitude of genes and are also structurally stable. They have specialized sequence elements in them that stabilize their structures and ensure their replication and segregation into daughter cells during cell division. In 1983, Murray and Szostak first described development of the yeast artificial chromosome (YAC) vectors [11], which were demonstrated to carry large DNA fragments by David Burke in 1987 [12]. DNA fragments larger than 100 kb and upto 3000 kb could be inserted and maintained in YACs as a linear yeast chromosome. YACs are cloning vectors derived from chromosomal DNA of the yeast *Saccharomyces cerevisiae*.

Similar to plasmids, YAC vectors are capable of replicating in bacteria using the Ori site. They also possess selection markers to allow enrichment of positive

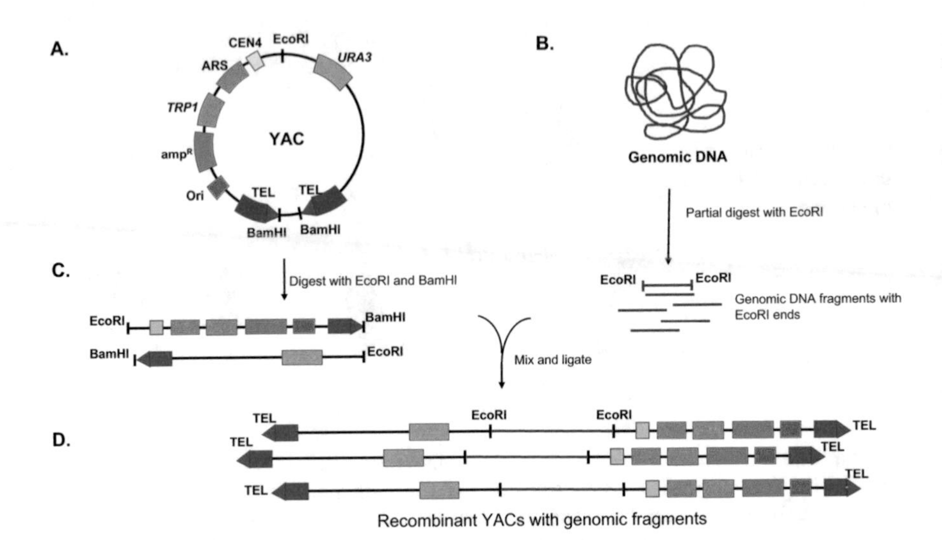

Fig. 2.7 Components of a YAC vector and steps of cloning in YAC. (**a**) YAC vector is digested using BamHI and EcoRI to divide it into two arms. (**b**) Genomic DNA is also digested with BamHI and EcoRI partially to get genomic DNA fragments with compatible ends for cloning into YAC. (**c**) Digested vectors and fragments are mixed together and (**d**) ligated resulting in recombinant YACs with genomic DNA fragments inserted into the YACs. The recombinant YACs are then transfected into suitable yeast strains

transformants. Being linear chromosomal structures, YACs require following features for successful replication in yeast cells (Fig. 2.7).

- Yeast autonomously replicating sequence (ARS1) which works as an origin of replication for eukaryotic yeast replication machinery,
- Yeast telomeres: Telomeres are specialized, repetitive sequence elements required for efficient replication of ends of linear chromosomal DNA molecules and to protect the ends of chromosomes from breakage. *Tetrahymena* telomeres have also been used to construct YAC vectors and are known to work efficiently in yeast.
- Yeast centromere: Centromere consists of AT rich sequence element that attaches to the spindle during cell division and ensures efficient segregation of chromosomes into the daughter cells. Presence of centromeric sequence in the chromosome maintains the chromosome copy number of one per cell.
- Selection markers for yeast: For selection of YAC transformants, certain yeast derived genes such as *ura*3 (orotidine 5-phosphate decarboxylase, an enzyme involved in de novo synthesis of pyrimidine ribonucleotides or *trp*1 (N-(5′-phosphoribosyl) anthranilate isomerase, an enzyme involved in tryptophan amino acid synthesis) are included in YACs. Transformants containing YAC vectors are capable of surviving in absence of the relevant nutrient source.

In addition to components required for replication and maintenance of the vector in yeast, YACs also possess bacterial origin of replication and selection marker genes to allow their propagation in bacterial cells before insertion of genomic DNA fragments. Cloning in YACs involves digesting the vector in two fragments and then ligating it with the foreign DNA fragment. Selection of recombinant YAC containing yeast clones is based on insertional inactivation as well as genetic complementation in mutant host cells. YACs contain selection marker genes that code for enzymes essential for metabolism of certain nutrients. These YACs are transfected into the yeast strains that have mutations in genes e.g., *ura3*, *trp1,* and a functional copy of these genes is provided by YAC resulting in reversal of metabolic defect.

2.3.2 *Bacterial Artificial Chromosome (BAC)*

In the 1990s, YACs were being used for implementation of the Human genome project. However, handling of yeast and YACs compared to *E. coli* and plasmids was challenging for molecular biologists. YACs could carry large fragments of human genome (upto 1 megabase), however, YACs had limitations. YACs were prone to forming chimeric clones (more than one genomic fragment inserted in a single YAC vector) making physical mapping of the genome difficult. Hence, to address these issues, BAC vectors were constructed using an *E. coli* F′ factor by Hiroaki and Shizuya in 1992. Bacterial F′ (fertility) factor is a plasmid that allows conjugation in bacterial cells where genes can be transferred from the donor bacterial cell (F positive) to the recipient cell (F negative) [13]. The F′ factor has a special Ori site and regulatory genes that regulate DNA synthesis allowing only a few copies (one or two copies) of the F′ plasmid per bacterial cell. For selection of transformants, BAC vectors are equipped with antibiotic resistance genes and lacZ' gene with restriction enzyme sites within it to allow insertional inactivation and blue-white selection.

P1-derived artificial chromosome (PAC) is a DNA vector derived from a combination of P1 bacteriophage genome and BAC. PACs contain low copy number origin of replication from P1 phage and are maintained at around one copy per bacterial cell. PACs can carry from 100 kb–300 kb insert DNA. PACs have antibiotic resistance genes as selection markers for transformants. Circularization of PACs depends on the Cre (Causes *recombination*) recombinase of the phage and the loxP (locus of crossover) sites present at the ends of P1 phage genome [14]. When the phage DNA is injected into the host cell, the loxP site through complementary sequence cross over and recombination by Cre enzyme causes circularization of the phage genome. PACs are also useful in genomic library construction and cloning of genomic fragments and genes. They offer ease in handling and manipulation as compared to YACs due to the bacterial host system.

2.3.3 *Human Artificial Chromosome (HAC)*

HACs were first constructed in late 1990s, to identify minimum essential sequence elements involved in the assembly and function of the centromere region of the human chromosomes. HACs are microchromosomes that resemble in structure with actual human chromosomes and are in the size range of 6–10 megabases. Endogenous human chromosomes were systematically reduced to recognize the minimal chromosomal units required for efficient segregation during mitosis. It was concluded that alpha-satellite DNA and portions of the flanking pericentric region were sufficient for centromere function and chromosome stability. HACs can also be created by de novo construction by assembling synthetic or cloned DNA fragments that provide chromosomal components. HACs were designed to overcome the problems faced with the YAC and BAC vectors such as unwanted integration and disruption of host genes and different expression levels compared to the host genome. HACs work as an extra chromosome that resembles in structure and composition to the host chromosomes thus acting as an ideal gene carrier for human cells.

2.4 Viral Vectors

Viral vectors are modified viral genomes that retain certain viral genes and also carry recombinant DNA fragments that are to be delivered into the host cells. Similar to bacteriophage-based vectors, mammalian viral genomes were also utilized as gene delivery vehicles. Viruses provide an advantage of targeting cells for efficient delivery of recombinant DNA. Thus, mammalian viruses were utilized to develop vectors for gene delivery into the mammalian cells not only in vitro but also in vivo and this opened tremendous possibilities in the field of recombinant DNA technology.

2.4.1 *Lentiviral Vectors*

Lentiviruses (Lentus, slow) are mammalian viruses characterized by slow growth due to long periods of latency and persistent infections in the host cells. In latent state the viruses remain dormant and do not replicate. Lentiviruses are known to integrate their genome into the host cell genome. Due to this ability of lentiviruses, they act as potential gene delivery vehicles. In 1989, lentiviral genome was first employed to carry and transfer a foreign gene into the host cell. Human immunodeficiency virus-1 (HIV-1) genome was modified to carry chloramphenicol acetyltranferase (CAT) gene in place of viral accessory nef gene. The purpose of these experiments was not to develop vectors but they initiated the development of lentiviral vectors.

Naldini et al. in 1996 developed the lentiviral vector system that consisted of three plasmids viz—the packaging plasmid, envelope plasmid and the transfer plasmid. Packaging plasmid was a mutant HIV-1 genome containing plasmid that expressed viral accessory proteins but could not package itself due to mutation. Cell tropism of the resultant virus was decided by the envelope plasmid which expressed the envelope protein and the transfer plasmid carried the transgene and the packaging signal that allowed packaging of transfer plasmid into the newly formed virions. The virus thus packaged carried the transgene that could be delivered to host cells. Second generation of lentiviral vector system used three plasmids, one of which expressed only those HIV-1 genes that were essential for viral replication (gag, pol, tat, and rev), envelope plasmid expressed envelope gene and the third plasmid contained transgene with its promoter and HIV-1 long terminal repeats on both sides. Presence of LTR made the expression of transgene dependent on transactivator (tat) protein. In the third generation lentiviral system's requirement of tat was abolished by replacing the LTRs with chimeric LTRs. The lentiviral vectors were modified from time to time to reduce the chances of generating replication competent viruses and thus to ensure safety. Lentiviral vector allows integration of the transgene into the host cell genome allowing stable expression of transgene. Figure 2.8 gives the details of the vector.

2.4.2 *Adenoviral Vectors*

Adenoviral vectors are based on Adenovirus, a non-enveloped virus that contains a linear, double-stranded genome. Two serotypes 1 and 5 are most common. The structural components of the virus comprise of three major proteins fiber, penton, and hexon, first two of which take part in virus binding to its receptors (coxsakie adenoviral receptor or CAR) on target cells and internalization into the host cells while the hexon protein makes nucleocapsid, that provide structure to the viral particle. First-generation vectors were created by replacing viral E1A gene with promoters such as CMV and by removing E3 gene. However, these vectors displayed considerable cytotoxicity. Second-generation vectors were created by removing viral E2 and E4 genes from the first-generation vector. The toxicity was reduced but it required specific cell lines stably expressing helper genes for packaging. Figure 2.9 represents the details of adenoviral vectors.

A summary table showing details of different types of cloning vectors, their advantages, and disadvantages is given below in Table 2.3.

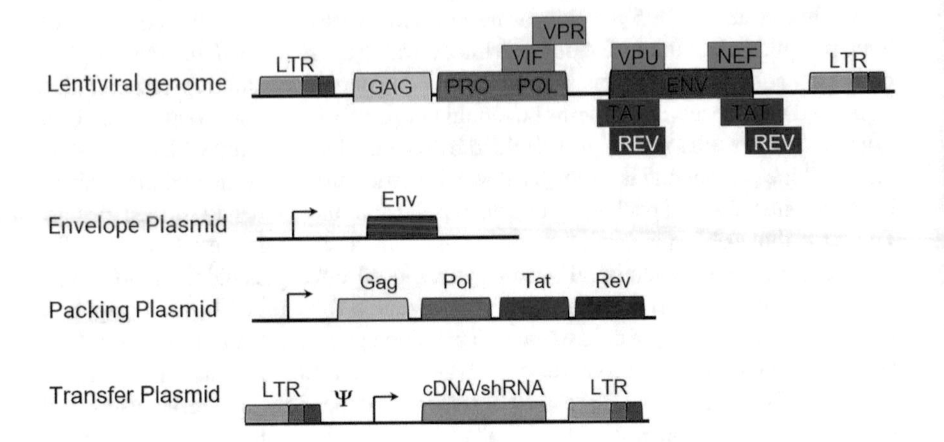

Fig. 2.8 The HIV-1 genome organization is shown on the top, from 5′ to 3′. The genome is separated A. Envelope plasmid, which expresses the required envelope glycoprotein. B. Packaging plasmid, which includes cytomegalovirus (CMV) promoter controlling Gag, Pol, Tat, Rev. C. Transfer vector (vector), that contains at least the LTR along with the promoter controlling the transcription of the gene of interest

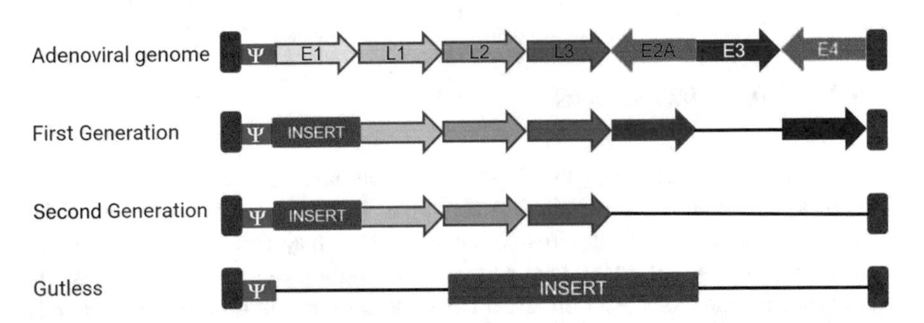

Fig. 2.9 The Adenoviral genome with essential elements is shown on top. The first-generation adenovirus vector is constructed by deleting the E1 and E3 regions from the adenoviral genome. The second-generation adenovirus has additional removal of E2 and E4 regions. The gutless adenovirus vector is constructed by deleting all the viral protein-coding genes, leaving only the ITRs and the packaging signal

2.5 Expression Vectors

Major applications of RDT can be seen in the field of recombinant protein production. Using RDT and cloning vectors, genes coding certain proteins of therapeutic or commercial importance can be expressed in bacterial, yeast, or mammalian host systems at a large scale. In order to promote expression of the protein the cloning vectors need to be equipped with the required components and such vectors are termed expression vectors. Expression vectors are plasmids designed especially for expression of the transgene in the desired host cell system. These vectors possess all

Table 2.3 Summary table of different types of cloning vectors

Sr. no	Type of vector	Insert size (kb)	Host for amplification	Advantages	Disadvantages
1	Plasmid	<10	Bacteria	Easy to manipulate, structurally stable, easy to transform	Low insert capacity, low transformation efficiency, stability reduces with size,
2	Phagemid	9–15	Bacteria	Increased insert capacity than plasmids, higher transformation efficiency, two Ori sites for phage and plasmid-based replication	Insert capacity restricted by phage head
3	Cosmid	23–45	Bacteria	Higher insert capacity, ease of manipulation, phage-based transformation	Formation of circular form during ligation prevents packaging Insert capacity limited by phage head capacity
4	BAC	100–300	Bacteria	Much improved insert capacity, easy handling similar to plasmid, maintain fragments stably	Low copy number leading to low yield, modification of insert through recombination is cumbersome
5	YAC	200–2000	Yeast	Very high insert capacity, a smaller number of clones required for genome coverage	Possibility of chimeric clones, fragile due to large size, difficult to separate from yeast chromosomes due to similar size
6	HAC	>2000	Human cell lines	Structurally and mitotically stable and remain as episome in human cells, low risk of insertional mutagenesis, safer than other vectors, have the capacity to carry and express large mammalian genes from their native gene regulatory elements	De novo synthesis of HACs is difficult as multiple, large genomic fragments do not get integrated easily into the vectors, delivery of HACs to the specific cell population is a challenge due to their large size

(continued)

Table 2.3 (continued)

Sr. no	Type of vector	Insert size (kb)	Host for amplification	Advantages	Disadvantages
7	Lentiviral vector	3–10	Mammalian cell lines and in vivo systems	Ability to transduce both dividing and non-dividing cells, stable, and high levels of transgene expression	Possibility of generation of replication competent viruses due to recombination, random integration of lentiviral vectors resulting in insertional inactivation of host genes, risk of oncogenesis due integration into the genome
8	Adenoviral vector	8–36	Mammalian cell lines and in vivo systems	High nuclear transfer efficiency, low pathogenicity of the virus, do not integrate into the host genome, they cause no disturbance to the host genes	Not suitable for long-term gene expression, works well only where high level transient gene expression is required

the basic components of the typical plasmid vector (Fig. 2.10). In addition to these components, expression vectors are equipped with other sequence elements that are necessary for the expression of cloned transgene. The goal of cloning a complementary DNA (cDNA) into the expression vector is to maximize the transcription and eventually translation of the cloned segment.

The components required for expression of transgene in a given host are

1. A strong promoter that ensures assembly of the transcription machinery for mRNA synthesis encoded by the transgene. Based on whether the final product of the transgene gene is RNA or protein the promoter is selected. For expression of small RNA elements such as shRNA, miRNA in mammalian systems human U6 promoter can be employed. Depending on whether the transgene expressed mRNA or small RNA, promoter sequence is chosen to ensure recruitment of respective RNA polymerase and transcription machinery. Traditionally, human cytomegalovirus (CMV) promoter was most widely used in the expression vectors. Better promoters have now been recruited to increase expression. Promoter of human elongation factor 1 alpha (EF1α promoter), Chicken beta actin and its versions are also used. Several synthetic promoter elements have also been constructed by combining features from different natural sequences. CAG is a strong hybrid mammalian promoter derived from combination of CMV enhancer sequence, chicken beta actin promoter and rabbit beta globin intron. Such promoter and enhancer sequences provide enhanced expression of transgene that is maintained at high levels for longer periods. Based on the host system e.g., bacterial, yeast, plant, or mammalian, the expression vectors are developed with appropriate components.

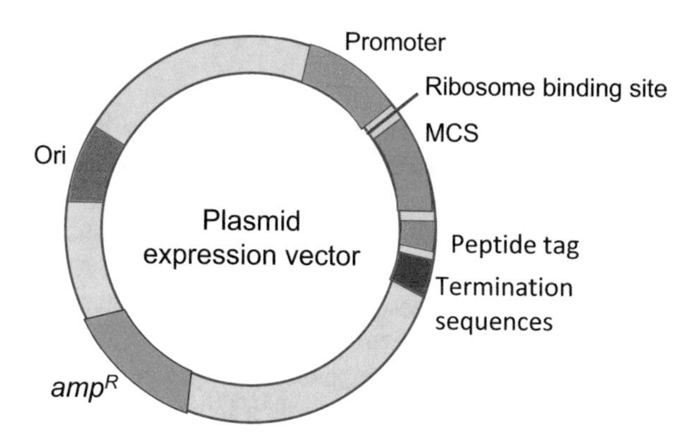

Fig. 2.10 Basic components of an expression vector. Promoter ensures expression of the transgene cloned within the MCS. Peptide tag is fused to the protein expressed by the transgene and is required for the purification of protein during downstream processing. Termination sequences are necessary to correctly stop the transcription of the transgene and eventually translation of mRNA. ampR is the selection marker and Ori is the origin of replication

2. A strong transcription termination sequence is also important to prevent the transcription machinery from running through the coding sequence. Similarly, ribosome entry sequence, a translation start site for assembly of ribosomes and other translation machinery and a stop codon is necessary for the protein translation machinery to recognize the end of translation frame.
3. Another important component in expression vectors used for production of recombinant protein is a protein tag. For purification of proteins expressed, tags are attached to proteins by adding them to the protein coding sequence while inserting it into the cloning vector. Protein tags consist of peptides that can be used for the purification of the recombinant protein by affinity column-based purification. These tags include 6X Histidine tag, HA or FLAG epitope/peptide tag.

Expression vectors are used for the expression of gene products. Specific mammalian expression vectors can be used to retain the folding, post translational modification and activity of the expressed protein in a mammalian system. Apart from expression of the gene product, the expression vectors also enable purification of the expressed protein. By varying the strength of gene regulators like promoters in the vector, the extent of gene expression can be modified. Bacterial expression systems pose challenges in expressing mammalian proteins. Mammalian expression systems involve a high cost in terms of culture media, transfection reagents, etc.

2.6 Shuttle Vectors

Cloning vectors, usually plasmids that have the ability to replicate in two different host species and can carry foreign DNA fragments from one host cell to the other, are termed shuttle vectors. Based on the two species, the cloning vector is equipped with the components required for replication and transgene expression and selection markers functional and specific to the two hosts. In most cases, one of the host species is bacteria. Shuttle vectors are designed to propagate in prokaryotic and eukaryotic hosts (e.g., in *E. coli* and in *Saccharomyces cerevisiae)* or two different bacterial species. Ability of shuttle vectors to propagate in two species makes their manipulation easier in the bacterial host while retaining their functionality in the yeast system which is difficult and slower for vector manipulation. A shuttle vector usually consists of an *E. coli* component and a yeast component. The *E. coli* components include the origin of replication and a selectable marker while the yeast component consists of the autonomously replicating sequence (ARS), a yeast centromere (CEN), and a yeast selectable marker. A typical example is YEp13, which is a *S. cerevisiae/E. coli* type yeast episomal vector.

Shuttle vectors are used for gene amplification, mutagenesis, and gene manipulation. They are used to transfer genes between different host cells. One of the major advantages of shuttle vectors is that it can be used in prokaryotes and eukaryotes. Genes cloned into these vectors can be manipulated in prokaryotes and used in a more complex system. However, the larger size of the vector may make transformation difficult in certain cases.

Examples of some of the commercially available vectors are given below for the benefit of the reader in Table 2.4.

2.7 Summary

Starting from plasmids, a variety of cloning vectors is available. Choice of cloning vector is mainly guided by the goal of the cloning experiment and factors including host cell type, insert size and vector size, restriction sites available for insertion of DNA fragment, Ori site, promoters and peptide tags in case of expression vector. Since the discovery of DNA as an inheritance material, the curiosity to understand how DNA rendered the phenotypes was born. Initially, simpler model systems like bacteria were utilized to carry out recombination experiments where genes from one bacterial species were transferred to another using plasmid to understand which phenotypic characteristics were transferred along. These experiments were further applied to higher organisms and cloning vectors played a particularly important role in genetic and genomic studies. RDT and cloning have revolutionized the field of molecular biology. The range of cloning vectors developed so far is vast and they make a basic tool for molecular biology techniques. Applications of cloning vectors in the health sector, agriculture, food, and other industries are unmatched.

Table 2.4 Some commercially available vectors

Vectors	Promoter	Selection marker	Source	Description
pASK75	tetA	amp^R	Biometra	Bacterial expression vector
pBAD series	araBAD	amp^R	Invitrogen	Inducible bacterial protein expression vector. The vectors have pUC Ori. araBAD promoter is derived from an operon which controls L-arabinose metabolism in *E. coli*. Gene of interest is cloned in pBAD vector downstream to araBAD promoter. araBAD promoter is activated in presence of L-arabinose and inhibited in presence of glucose allowing regulated expression of transgene. Other variants offer options of protein expression with various fusion proteins or cleavage sites
pET series	T7	amp^R/ kan^R	Novagen, Agilent	Widely used bacterial expression vector. 6xHis tag allows immobilized metal affinity chromatography mediated purification of protein. pET vectors are designed to obtain high level transcription and translation mediated by T7 bacteriophage gene. Plasmid versions are available that enable protein expression in all three possible reading frames relative to genes
pETM series	T7/lac	kan^R	MPI Biochemistry	These vectors are derived from pET vectors with various modifications for rapid production of proteins with specific yield, stability, and cleavage features
pGEX series	Tac	amp^R	Pharmacia, Sigma	Bacterial expression vectors containing GST tag and various options of protease cleavage sites such as Factor Xa, Thrombin and PreScission protease, etc
pProEx series	trc	amp^R	Life Technologies	Prokaryotic expression vectors with 6xHis tag and rTEV protease recognition site for efficient cleavage during purification
pQE series	T5/lac, p10, CAG	amp^R	Qiagen	Vector for expression of his-tagged proteins in bacterial, mammalian, or insect systems using a single vector
pEGFP-series	CMV	kan^R, neo^R	Addgene, Takara Bio	Vector for visualization of protein of interest expressed as a fusion protein with GFP
pcDNA series	CMV	amp^R, neo^R	Thermo Fisher, NovoPro	Mammalian expression vector
pLKO.1-puro	U6	amp^R, $puro^R$	Sigma	Replication incompetent lentiviral vector used for expression of shRNA and production of lentiviral particles

(continued)

Table 2.4 (continued)

Vectors	Promoter	Selection marker	Source	Description
pGL3 series	Based on versions	*amp*^{R}	Promega	pGL3 series of vectors are reporter vectors used for studying the regulation of mammalian gene expression through quantitative assays. pGL3-basic vector has luciferase gene without a promoter. One can clone a promoter of interest upstream to the luciferase gene to study promoter regulation. Similarly, other vectors available in this series allow analysis of enhancer sequences

- Plasmids are circular, self-replicating double-stranded DNA molecules that are inherited from one generation to another in bacteria. Plasmids have been modified to serve as cloning vectors for application in RDT. Plasmids offer an easy to manipulate and propagate cloning vector that can be amplified in bacterial systems. Multiple versions of plasmid vectors have been created for various purposes.
- Bacteriophage-based vectors including lambda phage-based vectors, cosmids, M13 phage-based vectors and phagemids have been developed using bacteriophage genomes or their components. Phage-based vectors provide increased insert carrying capacity along with higher transformation efficiency through infection of bacterial cells.
- Artificial chromosome vectors were eventually developed to further improve insert carrying capacity of the vector. Eukaryotic chromosome-like features were included in the vector to mimic chromosome structure. These vectors could carry very large inserts and still maintained stability.
- Cloning vectors, mainly plasmids, were further modified for special purposes such as expression of transgene. Expression vectors were equipped with sequence elements necessary for efficient expression of transgene promising high protein or RNA yield.

Currently, cloning vectors are being exploited for their potential in gene therapy and other therapeutic strategies. The major limitation vectors face is that of delivery to the intended tissue or cells. Vectors as DNA molecules are vulnerable to degradation by the host defense mechanisms. They need to be protected by viral capsids or lipid encapsulations to allow them to be delivered inside the target cells. Hence, development of cloning vectors that can overcome these hurdles faced during their delivery is a need of time.

Self Assesment

Q1. What factors need to be considered while selecting the vector for a specific DNA cloning experiment?

A1. Multiple factors should be considered carefully before planning a strategy for a cloning experiment.

1. *Purpose of the cloning experiment*: Choice of vector will be mainly guided by the reason why a particular cloning experiment is performed. For example, when a protein is to be expressed in mammalian cells, an expression vector equipped with a eukaryotic promoter is required. Similarly, the vector should be able to replicate in bacterial system for manipulation and amplification and should also be able to propagate and express in mammalian cells.
2. *Insert size*: Vectors vary in their insert carrying capacity should be chosen accordingly.
3. *Restriction sites*: The selected vector should have suitable restriction sites used for insertion of the DNA fragment. The sites should be unique in a way that they are not present in the insert sequence and should be present only in the MCS region of the vector. Thus, during digestion, the vector and the insert are cut only at expected sites.
4. *Copy number*: Replication efficiency of a vector varies with bacterial strain used and the Ori site they possess. Depending on the need of the experiment a vector with high or low copy number needs to be selected. For example, vector with low copy number is required for generation of genomic libraries to prevent recombination in the original genomic fragment sequence. On the contrary, vector with high copy number is selected for overexpression of a protein, or where high insert DNA yield for sequencing is intended.
5. *Selection marker*: Depending on host cell system used in an experiment, available resources, a vector with appropriate selection marker is chosen. For example, a selection marker gene is required for selection of transformants. In a bacterial host system a selection marker responsible for resistance to antibiotic such as ampicillin or kanamycin would be used, whereas for mammalian cell system an resistance gene for antibiotic suitable for eukaryotic cells such as neomycin is necessary. Sometimes, vectors are chosen based on available resources for detection or screening of transformants. A vector with fluorescent protein expression as a selection marker can be chosen when fluorescence detection and visualization instrument like a fluorescence microscope is available.
6. *Downstream processing steps*: Carefully chosen vector expressing a tag peptide fused to the expressed protein would lead to easy affinity purification.

Q2. A student is working on a protein involved in nuclear-cytoplasmic trafficking in mammalian cells. The student wants to express that protein in mammalian cells and study its subcellular localization in different growth conditions in live cells. Help the student to choose a vector for the experiment.

A2. Since the host system in this experiment is mammalian cells, a vector suitable for mammalian cells is required. The student wants to express the protein of interest and study its subcellular localization, hence a eukaryotic expression vector with a peptide/protein tag that will allow visualization of protein in live cells is required.

A plasmid vector like pEGFP-N1 can be used which is designed for expression in mammalian cells with CMV promoter and has a GFP tag included in the vector expressing the protein of interest as a fusion protein (GFP fused with the protein of interest) allowing its visualization in live cells.

Q3. A researcher has cloned a DNA fragment (~1.8 kb) in pcDNA3.1+ vector (5446 bp) vector at BamHI site. Following transformation into *E. coli*, the researcher needs to identify the clones (bacterial colonies with the vector in them) with vectors containing the insert in the correct orientation. The researcher did not use the *XbaI* site as it was present in the insert at about 1.5 kb from the 5′ end of the insert. Can you propose a strategy to identify the right clones?

A3. The correct clones can easily be identified with the use of specific restriction enzymes. The clones that grow on the selection medium containing specific antibiotic medium may contain recombinant vectors in three forms. The screening for correct recombinant plasmid can be done by digestion of the plasmids using XbaI as shown in table below. The clones are identified based on the migration pattern of Xba1 digested plasmid bands on an agarose gel.

Scenario	Vector	Band pattern on the agarose gel
1.	Empty vector without insert	Single linearized vector band corresponding to size of the vector
2.	Recombinant vector with insert in reverse orientation	Two bands of 5.7 kb (5.4 kb $_{vector}$ + 300 bp $_{insert}$) and 1.5 kb $_{insert}$ with two XbaI sites on either end. The clones will be digested at two XbaI sites one in the insert and other in the vector. However, due to reverse orientation of the insert two XbaI sites are separated by 1.5 kb
3.	Recombinant vector with insert in correct orientation	Two bands of 6.9 kb (5.4 kb $_{vector}$ + 1.5 kb $_{insert}$) and 300 bp $_{insert}$ with two XbaI sites on either end The clones will be digested at two XbaI sites one in the insert and other in the vector. Correct orientation of the insert two XbaI sites are separated by 300 bp

Q4. A researcher plans to study RNAi-based suppression of a target protein expression. His experimental design involves creation of a luciferase-based reporter system in a mammalian cell line for the experiment. He plans to express candidate shRNAs (small hairpin RNAs) in the reporter system that expresses target protein fused with luciferase. Thus, shRNA mediate suppression of protein expression can be monitored by reduced luciferase levels. His shRNA construct, pKLO.1-puro vector contains two Ori—pUC and SV40 and a U6 promoter. Can you explain why he chose these plasmids with two Ori and what is the role of U6 promoter present in this vector?

A4. Two different Ori help the plasmid to replicate in different host organisms and are termed as shuttle vectors. pUC Ori is the bacterial origin of replication. The vector can be amplified in a bacterial host for manipulation. SV40 Ori is functional in the mammalian cells and used for plasmid replication in the eukaryotic cells.

In human cells, U6 promoter is responsible for expression of small nuclear RNAs by recruitment of type III RNA polymerase. For this experiment, the production of shRNA will be under the control of U6 promoter.

Q5. A researcher has a YAC with three Ori sites—ARS, f1 Ori, and pBR322 Ori. The researcher intends to use the YAC in *E. coli* and Yeast. Do you think that the f1 Ori will affect the vector usability in any of these hosts? What is the role of f1 Ori?

A5. No. The researcher is using a shuttle vector wherein pBR322 Ori is a bacteria Origin of replication and ARS works in yeast. f1 origin is derived from filamentous phages that allow replication of single-stranded DNA to be packaged in the phage head. In absence of helper phage f1 ori is totally nonfunctional, inert, and harmless.

Q6. What is a Kozak sequence and is it necessary for protein expression? A researcher plans to express a protein tyrosine kinase in mammalian cells using pcDNA3.1 vector. Is it necessary to incorporate a Kozak sequence before the ATG initiation codon to ensure the expression?

A6. The Kozak sequence (A/GCCAUGG) in mammalian mRNA enhances translation from correct initiation codon. it is required but not necessary. pcDNA contains a strong CMV promoter that would have high expression of mRNA. In addition, the presence of SV40 polyA provides enhanced stability of mRNA. Therefore, in absence of Kozak sequence, the researcher may still obtain a good translation rate through pcDNA vector's strong mRNA expression. However, it would be a good idea to include 5′ Kozak sequence in the same vector to have high protein expression.

References

1. Lederberg J. Cell genetics and hereditary symbiosis. Physiol Rev. 1952;32(4):403–30.
2. Lodish H, Berk A, Zipursky SL, Matsudaira P, Baltimore D, Darnell J. Molecular cell biology. 4th ed. W. H. Freeman; 2000.
3. Cohen SN, Chang ACY, Hsu L. Nonchromosomal antibiotic resistance in bacteria: genetic transformation of Escherichia coli by R-factor DNA*. Proc Natl Acad Sci U S A. 1972;69(8):2110–4.
4. del Solar G, Giraldo R, Ruiz-Echevarría MJ, Espinosa M, Díaz-Orejas R. Replication and control of circular bacterial plasmids. Microbiol Mol Biol Rev. 1998;62(2):434–64.
5. Vieira J, Messing J. The pUC plasmids, an M13mp7-derived system for insertion mutagenesis and sequencing with synthetic universal primers. Gene. 1982;19(3):259–68.
6. Feiss M, Widner W. Bacteriophage lambda DNA packaging: scanning for the terminal cohesive end site during packaging. Proc Natl Acad Sci U S A. 1982;79(11):3498–502.
7. Collins J, Brüning HJ. Plasmids useable as gene-cloning vectors in an in vitro packaging by coliphage lambda: "cosmids.". Gene. 1978;4(2):85–107.
8. Hohn B, Murray K. Packaging recombinant DNA molecules into bacteriophage particles in vitro. Proc Natl Acad Sci U S A. 1977;74(8):3259–63.

9. Messing J, Gronenborn B, Müller-Hill B, Hans Hopschneider P. Filamentous coliphage M13 as a cloning vehicle: insertion of a HindII fragment of the lac regulatory region in M13 replicative form in vitro. Proc Natl Acad Sci U S A. 1977;74(9):3642–6.
10. Bazan J, Całkosiński I, Gamian A. Phage display—a powerful technique for immunotherapy. Hum Vaccin Immunother. 2012;8(12):1817–28.
11. Murray AW, Szostak JW. Construction of artificial chromosomes in yeast. Nature. 1983;305(5931):189–93.
12. Burke DT. The role of yeast artificial chromosome clones in generating genome maps. Curr Opin Genet Dev. 1991;1(1):69–74.
13. Shizuya H, Birren B, Kim UJ, Mancino V, Slepak T, Tachiiri Y, et al. Cloning and stable maintenance of 300-kilobase-pair fragments of human DNA in Escherichia coli using an F-factor-based vector. Proc Natl Acad Sci U S A. 1992;89(18):8794–7.
14. Bajpai B. High capacity vectors. In: Advances in biotechnology; 2013. p. 1–10.

Chapter 3
Gene Isolation Methods: Beginner's Guide

Rajendra Patil, Aruna Sivaram, and Nayana Patil

3.1 Introduction

Recombinant DNA technology is the field that encompasses all the techniques used in artificial modification of organism's DNA for production of desired product or to increase/decrease the expression of genes for industrial, agricultural, or medical applications. The first step in this process is the isolation of a target gene from the organism which starts with extraction and purification of the DNA. This process is followed by desired manipulation of the source DNA using various laboratory techniques. The manipulated DNA is then inserted in the suitable genetic carrier and delivered to the host of interest [1]. For any genetic modification the imperative step remains isolation of desired genes. Current chapter will be fixated on the tools of gene isolation. Moreover, use of bioinformatics tools and techniques has shed the burden off in genetic engineering. It has made the prolonged and cumbersome tasks in genetic engineering easier and time efficient. Bioinformatics has opened the doors to the technology where the information can be analyzed and precise data can be generated with lesser time requirement [2]. This chapter will also try to highlight the bioinformatics tools that can be efficiently applied for the isolation of the gene.

The prerequisite for cloning a gene is getting high quality and pure nucleic acids from organisms of interest using appropriate protocols. The extraction method may differ depending on the sources but involves steps optimized to get high yield and

R. Patil
Department of Biotechnology, Savitribai Phule Pune University, Pune, Maharashtra, India
e-mail: rpatil@unipune.ac.in

A. Sivaram · N. Patil (✉)
School of Bioengineering Sciences & Research, MIT ADT University,
Pune, Maharashtra, India
e-mail: aruna.sivaram@mituniversity.edu.in; nayana.patil@mituniversity.edu.in

N. Patil, A. Sivaram, *A Complete Guide to Gene Cloning: From Basic to Advanced*, Techniques in Life Science and Biomedicine for the Non-Expert,
https://doi.org/10.1007/978-3-030-96851-9_3

quality DNA or RNA to ensure it is well suited for enzymatic reaction in the consecutive steps. The entire process of DNA isolation utilizes the physio-chemical properties of the DNA molecules such as charge, polar nature, hydrophilicity, etc. to separate it from contaminants such as lipids, protein, polysaccharides, cell debris, etc. To further facilitate the extraction procedure, several enzymes (RNA, lysozyme, proteases) and reagents (salts, detergents, organic solution) are also employed.

3.2 Preparation of Genomic DNA

There are various physical, chemical, and enzymatic methods to obtain DNA from cells [3]. Irrespective of method used or the source of DNA, all the steps are selected and optimized to get maximum quantity, quality, and integrity of genomic DNA. As DNA is part of cytoplasm and is surrounded by nuclear and or cell membrane therefore prior to nucleic acid purification, elimination of cell debris, lipids, and proteins is carried out in three important steps (Fig. 3.1): disruption of cell membrane/cell wall, separation and precipitation of DNA, and lastly dissolving DNA.

3.2.1 Disruption of Cell Membrane/Cell Wall

The lysis of prokaryotic cells and animal cells is achieved by using hypotonic shock, pH-based permeability, enzymatic degradation, or detergents [3, 4]. The plant cell lysis involves an additional step of mechanically breaking the cells by grinding using mortar pestle [4].

Hypotonic shock: The salts used in solution-based lysis of cells leads to hypotonic shock inducing swelling of the cells making it fragile and easy to burst open. Tris present in the lysis buffer interacts with the lipopolysaccharides and makes

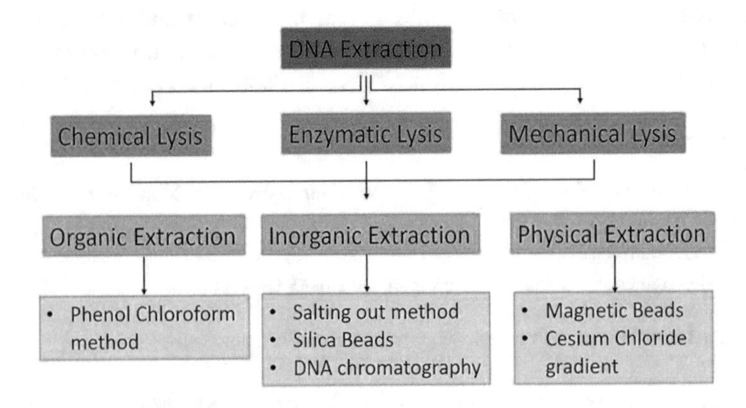

Fig. 3.1 Different types of DNA isolation methods

them permeable. The enzymes such as lysozyme attack the cell wall present in bacterial cells making it prone to lysis. Ionic detergents and surfactants solubilize the cell and nuclear membrane and help the membrane to break open.

Upon lysis of a cell, all the content present in the cells such as proteins, lipids, RNA, polysaccharide, and enzyme leaks out. Most of these can damage the integrity of DNA or may interfere with DNA precipitation. Hence, chelating agents such as EDTA are used which sequesters Mg2+ cofactor leading to inactivation and blocking of DNAse enzymes and hence protecting DNA from getting degraded. Another set of enzymes such as RNA and Proteases are used to digest and to get rid of RNA and protein contaminants. Chemicals such as SDS, β-mercaptoethanol, Triton, and CTAB mask the native charges of the proteins and destabilize its structure, making it easy to precipitate in the consecutive steps.

3.2.2 Separation and Precipitation of DNA

The separation of DNA from the other soluble macromolecules and insoluble debris is achieved by organic phenol–chloroform method and salting out treatment [5].

Organic extraction: The difference in the polarity between DNA and organic solvents helps to fractionate nonpolar components of the cells in the organic phase while the DNA remains in the aqueous phase. The addition of chloroform to the mix enhances the solubility of nonpolar cell debris, denatured proteins and lipids in phenol promoting the partition of contaminants in organic phase while DNA being polar tends to remain in aqueous phase. Additionally, the solubility and stability of DNA in aqueous phase is enhanced by binding of Na^+ cation to negatively charged phosphate backbone. The organic and aqueous phases are removed by centrifugation or vacuum filtration. The dissolved DNA can be fetched by salting out the DNA. The collected aqueous phase containing the dissolved nucleic acid is precipitated by using monovalent salts such as sodium chloride, potassium acetate, sodium acetate or ammonium acetate. The positively charged ions of the salt bind and neutralize the charges on sugar phosphate backbone allowing the DNA to agglomerate. The DNA–salt complex is precipitated using alcohol. The low dielectric constant of ethanol or isopropanol reduces hydrophilicity of DNA hence causing it to precipitate out of the solution. The precipitate is recovered by centrifugation, the pellet obtained is washed twice to get rid of the salt and finally the DNA dissolved in the appropriate buffer.

3.2.3 Silica-Based Extraction

After lysis of cells (as described in Sect. 3.2.1), the negatively charged DNA present in the cell lysate interacts with the positively charged silica membrane and is retained on the membrane while the rest of the unbound soluble material filters through the membrane. The impurities and loosely bound molecules are removed using a washing buffer. In the last step, the DNA bound is eluted by dissolving it in a buffer and collecting. The method is made time efficient by developing a silica-based spin column compatible with centrifuge or vacuum filtration systems [6].

3.2.4 DNA Chromatography

In this method, positively charged resin such as styrene-divinylbenzene copolymers, diethylaminoethyl is filled in column to form an anion exchanging matrix. The cell lysate containing genomic DNA is allowed to pass through the stationary phase, allowing the DNA to reversibly bind to positively charged resin. After getting rid of the impurities in the washing step, the bound DNA is eluted with a high salt buffer and precipitated using alcohol [3, 6].

3.2.5 Magnetic Beads

Similar to anion exchange chromatography, positively charged magnetic beads entrap negatively charged DNA. The DNA loaded beads are concentrated and separated from the solution using a magnetic field. The DNA can then be precipitated using ethanol as described earlier [3, 6].

3.2.6 Cesium Chloride Method

Instead of using charge on the DNA, molecular weight of the DNA can also be used to extract DNA from cell lysate [7]. A density gradient of CsCl is established in a centrifuge tube, cell lysate is loaded on it and centrifuged at high speed. The DNA settles down as a band in the gradient where the density of DNA and gradient is the same.

3.3 From Genome to Gene of Interest

Once high quality and pure genomic DNA is obtained, the next task is to find out the gene of interest from the millions of base pairs present within the genome. Hence, locating and isolating genes is a big task especially when the sequence is not known. The process of isolating and cloning genes can take two approaches based on the prior knowledge of its sequence.

- If the sequence is not available, then the whole genome is fragmented, cloned to produce a genome library and then target gene is identified by searchable criteria such as sequence, protein product or enzyme activity.
- On the other hand, if the sequence of the gene of interest is available, it is much easier to amplify it using PCR methods from source DNA or artificially synthesis and followed by cloning in a suitable vector.

3.3.1 Genome/Gene Library

If the genome sequence is not available in such cases the genomic DNA is randomly fragmented, each fragment is cloned and screened for the phenotype. The collection of all the clones carrying at least one fragment is termed a whole genome library [8]. Breaking of the DNA molecules by sheer mechanical forces is possible but precise cutting of the DNA for obtaining a particular gene is a cumbersome task. This challenge is overcome by the discovery of the enzymes known as restriction enzymes (restriction endonucleases). These are the enzymes responsible for recognizing and cutting the DNA double helix at specific sites thus cleaving the DNA molecule into the fragments of precise size and compatible ends. Discovery of these enzymes revolutionized the methods of gene isolation. Restriction enzymes/Restriction endonucleases (RE) are derived from the bacteria where they provide protection to the bacterial DNA from the intruding foreign viral DNA by degrading it. They have a vital contribution in the physical mapping of the DNA. Sequence specificities of different RE are distinct, allowing each of the RE to recognize a specific sequence of four to eight nucleotides in DNA known as palindrome sequence for cutting the DNA. The restriction digestion reaction of the DNA is carried out in condition which leads to only partial digestion generating overlapping fragments.

The insertion site in the vector DNA is also prepared by digesting the vector with the same restriction enzyme. The compatible ends generated allow the DNA fragment to ligate with the vector DNA. Post ligation the recombinant plasmid generated is subsequently transformed into a host cell for amplification. Figure 3.2a gives an overview of construction of genomic DNA library.

3.3.2 cDNA Library

If the insert DNA belongs to a prokaryotic organism, it's feasible to fragment the whole genome and insert into a vector creating a gene bank or library. However, in eukaryotes as the gene is composed of noncoding introns, regulatory regions and large in size, for cloning functional gene usually RNA is reverse transcribed to complementary DNA (cDNA) and then inserted into a vector generating cDNA library [9]. A cDNA library represents cloning of only coding mRNA populations through a cDNA intermediate.

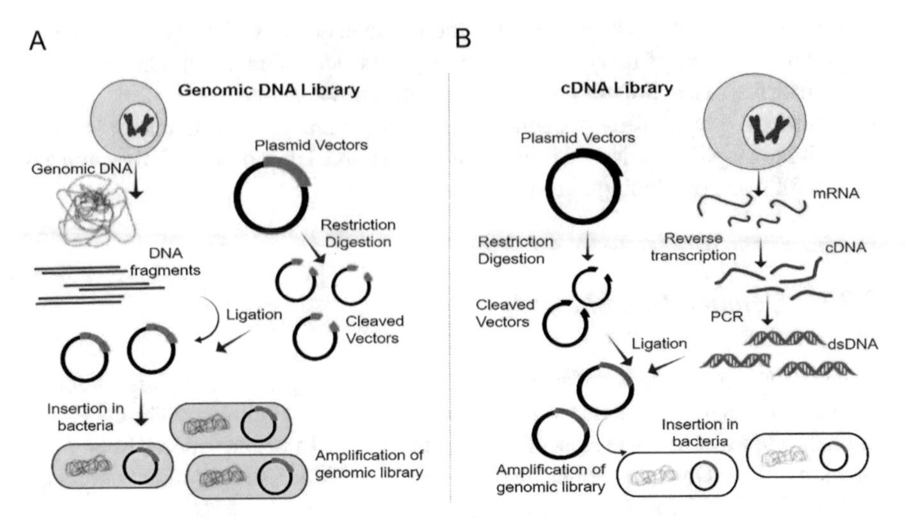

Fig. 3.2 (**a**) Overview of creating gDNA Library: The genomic DNA isolated from cells and suitable vector plasmid are digested with a restriction enzyme. The digested genomic DNA and the vector are ligated together and transformed into bacterial host cells. (**b**) Schematic representation of the cDNA library construction: mRNA is isolated from cells and reverse transcribed to get complementary DNA molecules. Double-stranded DNA is obtained by PCR amplification of cDNA. The resulting fragments are cloned in a suitable vector and inserted into bacterial host cells

Table 3.1 Differences between genomic and cDNA library

Genomic library	cDNA library
Made from total genomic DNA	Made from mRNA
It contains gene and regulatory regions	It contains only gene sequences
Gene cannot be expressed in cloning host	Gene can be expressed in cloning host
DNA insert is obtained by using restriction enzymes	DNA insert is obtained by using reverse transcriptase
All genes are present at the same frequency in a genomic DNA library	Frequency of a gene in a cDNA library depends on the abundance of corresponding mRNA
Collection of genes does not depend on cells or tissue used to isolate DNA	Collection of cloned gene sequence depends on cells or tissues used to isolate mRNA
Difficult to identify gene of interest due to large number of recombinants	Relatively simple to identify gene based on protein expressed
It will have clone of every gene for an organism	It will have clone of every gene for particular cell or tissue.
Genomic libraries are preferred for small genomes such as prokaryotic organisms	cDNA libraries are preferred for larger genomes as only exons are cloned
Useful for genome sequencing, analysis, and promoter studies	Useful for analysis of coding region

Differences between genomic and cDNA libraries are given in Table 3.1.

Figure 3.2b gives an overview of construction of cDNA library. The process of creating a cDNA library starts with isolation of mature mRNA from the cell. The mRNA is purified using the poly-T affinity column where only the mRNA possessing poly-A tail is retained in the column while the rest of the RNA molecules are washed out. The bound mRNA is then eluted using an appropriate elution buffer. A poly dT oligonucleotide is used as a primer by reverse transcriptase to synthesize complementary DNA strand forming DNA/RNA duplex. The RNA is removed from the duplex by using RNase H hydrolytic endonuclease leaving behind a single-stranded DNA molecule. In the last step, the ssDNA is converted to double-stranded DNA by employing DNA polymerase and the resulting cDNA is then inserted into the chosen vector and transformed into a suitable host.

3.3.3 Chromosome Walking

Once the genome library or cDNA library is constructed, in the next step, the clones are screened for specific genes of interest. Screening to identify the gene most of the time is based on DNA sequence homology, the protein product, or enzyme activity; each type of these methods are described in Chap. 7.

However, if the gene sequence is not known or the gene does not have any measurable phenotype, in such situations Chromosome or Genome walking is one way to deduce the gene sequence [10]. Chromosome walking relies on the closest known DNA sequence (usually a conserved sequence) termed molecular marker flanking target gene whose sequence is to be determined.

Chromosome walking begins with one clone from a library, and then identifies a second clone whose insert overlaps with the insert in the first clone. This process is repeated several times to walk across the DNA fragment and reach the gene of interest. A restriction fragment isolated from the end of the positive clones is used to reprobe the genomic library for overlapping clones.

Chromosome walking procedure:

1. From the genomic library a DNA fragment containing a marker gene is selected.
2. The fragment is digested with a series of restriction enzymes and electrophoresed to obtain a pattern i.e., a restriction map.
3. A small fragment from one end of the restriction map is subcloned. This end fragment serves as a probe to detect the next overlapping fragment.
4. Each probe is used as a primer to synthesize the next short DNA strand and the process continues till the end of the DNA fragment or till the unmapped gene in question is identified by DNA sequencing.

DNA Sequencing

DNA sequencing is the tool used for determining the nucleotide sequence of the nucleic acid. This sequence gives information about the gene or the genome. There are different methods [11] used for DNA sequencing that are mentioned below:

Sanger was the first to describe the method of DNA sequencing based on the selective incorporation of chain terminating dideoxynucleotides by DNA polymerase during vitro DNA replication. The Maxam-Gilbert method was based on nucleobase-specific partial chemical modification of DNA and subsequent cleavage of the DNA backbone at sites adjacent to the modified nucleotides. These two methods are considered as the basic methods of sequencing. The advanced method includes use of shotgun sequencing used for the long strands of DNA.

Next generation DNA sequencing includes Illumina sequencing which is based on reversible dye-terminators that enable the identification of single bases as they are introduced into DNA strands. Another one is the Pyrosequencing method that relies on the "sequencing by synthesis" principle, here the sequencing is done by detecting the nucleotide incorporated by a DNA polymerase. Solid-state (and biological) sequencing method is another method developed for sequencing which is a nanopore-based DNA sequencing method aiding in fast and high-resolution recognition and detection of DNA bases.

Advantages
- Unknown sequences of chromosomes can be identified by probing the nearest known sequence.
- It is a useful tool to track regions upstream and downstream of a marker sequence.

Disadvantages
- As the sequence is determined piece by piece, chromosome walking is time consuming and tedious.
- If the marker and gene of interest are far apart, using chromosome walking is not a practical approach.
- Unclonable DNA sections and repetitive DNA regions act as a barrier.

3.3.4 PCR Amplification of Target Gene

Polymerase chain reaction (PCR) is a technique to isolate genes of known sequences. For cloning a gene, multiple copies of the target DNA are required which is achieved by DNA amplification [12]. The length of the target sequence required for PCR procedure ranges between few base pairs to kilobases. For performing the PCR template DNA (genomic DNA, cDNA, plasmid) is amplified using two primers which are less than 30 nucleotides in length. They anneal to the template to be amplified at the beginning and end of it. This is the site where the DNA polymerase will bind and initiate the synthesis of new strands of DNA. The complementary strand is synthesized using DNA polymerase isolated from an extremophilic bacterium, *Thermus aquaticus*. Being isolated from thermophile Taq polymerase can withstand temperature up to 100 °C and several rounds of PCR reaction without losing its activity. DNA polymerase uses four deoxyribonucleoside triphosphates (dNTPs) as building blocks to synthesize the complementary strand during PCR reaction. The PCR process generally consists of 20–35 cycles and proceeds in three steps, viz. denaturation, annealing, and extension.

Denaturation: In order to use the double-stranded DNA as a template, the strands need to be separated exposing the sites where primers can bind. As the DNA strands are kept together in a helix by hydrogen bonds, raising the temperature to 95 °C for

a short time period breaks the weak hydrogen bonds, allowing the strands to separate.
Annealing: The PCR reaction is rapidly cooled down allowing the hydrogen bonds to reform. The primers hybridize more easily due to its higher concentration and short length (20–30 nucleotides) than long strands of substrate DNA. The annealing temperature called primer hybridization temperature is optimized for more selective and specific hybridization.
Extension: In the next step, the temperature of reaction is raised to attain optimal temperature for DNA polymerase to recognize the primed single-stranded DNAs and replicate the substrate DNA.

Box 3.1
Various modifications in PCR techniques are now available as multiplex PCR, Nested PCR, RT-PCR (reverse transcription), Hot-Start PCR, Long-Range PCR, qPCR, methylation specific PCR, Multiplex PCR, etc. There are certain limitations with the use of PCR as they require costlier equipment, more chances of contamination and its sensitivity toward some classes of contaminants and inhibitors, the thermal cycling required, etc. These limitations demanded the birth of alternative methods and gave rise to techniques such as nucleic acid sequence-based amplification (NASBA), loop mediated isothermal amplification (LAMP), rolling circle amplification (RCA), self-sustained sequence replication (3SR), etc. Many of these alternative techniques are isothermal nucleic acid amplification methods thus obviating the requirement of thermal cycler. The alternatives for the PCR methods are well reviewed by Fakruddin et al. [13]. Although there are raising methods for gene amplification PCR is still widely and commonly used for isolation of the gene.

Advantages

It can amplify minute amounts of template DNA.
As primers anneal to short sequences in DNA, DNA samples of poor quality can still be analyzed.
Cells which take long period to grown or which are difficult to culture in vitro can be studied using PCR.
PCR is rapid in providing results.

Disadvantages

PCR depends on primers for amplification hence can only amplify known and specific targets.
PCR product size is very limited, only fraction of the genome can be amplified at a time.
DNA polymerase can introduce errors while amplification of DNA.
PCR does not provide information about quantity of a target sequence.
The test cost and equipment cost is high.

3.3.5 Artificial Synthesis of DNA

Artificial DNA synthesis has become a routine technique to obtain short fragments of DNA used as specific primers, probes or gene editing templates. However large fragments of gene, entire chromosomes, sequences that are hard to isolate or even nonnatural genes are synthesized using Chemical [14] and enzyme-based methods [15]. Figure 3.3 gives a schematic representation of artificial DNA synthesis.

Chemical or Phosphoramidite Method

Phosphoramidite synthesis method can build an oligonucleotide up to 300 base pairs. The assembly of nucleotides needs a solid base usually controlled pore glass (CPG) or macroporous polystyrene (MPPS) which provides solid support to the growing chain of oligonucleotide. The synthesis occurs in four steps adding one base at a time to the preexisting oligonucleotide chain allowing it to grow by one unit at the end of each cycle.

Step 1: The nucleotides used in phosphoramidite synthesis are usually protected by dimethoxytrityl (DMT), acid treatment is used to deprotect the nucleotides. The acids such as trichloroacetic acid (TCA) react with DMT exposing the 5′-hydroxyl groups of nucleotides by this means making it available for attachment of the next nucleotide.

Step 2: The next DMT-protected base is added which attaches to 5′-OH groups of previous nucleotides forming a phosphite triester.

Step 3: In order to reduce error or deletion in the desired oligonucleotide sequence, 5′-OH group that has not be used for extension are capped by acylation using acetic anhydride.

Step 4: In the last step, the unstable phosphite triester bonds connecting the monomers are oxidized using iodine solution forming a stable cyanoethyl-protected phosphate backbone.

The reaction steps are repeats for the next base in the same sequence till the desired sequence has been synthesized. The product obtained at the end is detached

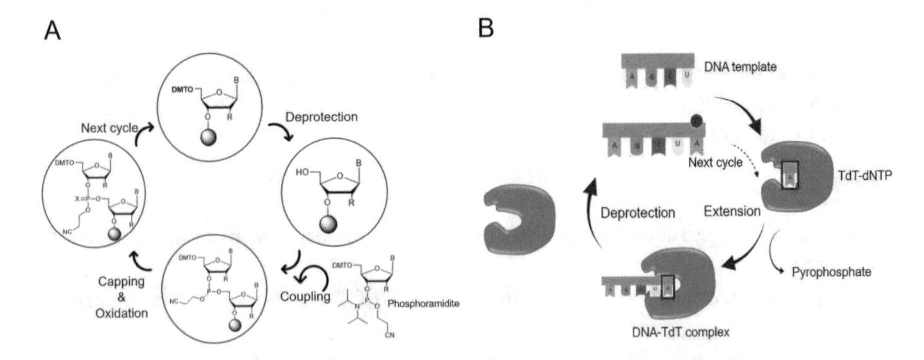

Fig. 3.3 Reaction cycle in (**a**). Standard automated solid-phase phosphoramidite-based oligonucleotide synthesis. (**b**) TdT enzyme-based oligonucleotide synthesis

from the CPG support and dimethoxytrityl molecule on the synthetic DNA is removed.

Advantage
- It is simple and highly efficient.

Disadvantage
- It gives low yield.
- The longer coupling reaction is time consuming, as well as the chemical reagents used in the synthesis process leads to generation of toxic waste. Several incorrect interactions occur leading to some defective products.

Enzyme-Based or Terminal Deoxynucleotidyl Transferase (TdT) Method
To overcome the drawbacks of chemical-based synthesis, enzymatic DNA synthesis is a promising alternative. The use of enzymes gives more specificity, large product size synthesis under mild conditions.

Terminal Deoxynucleotidyl Transferase (TdT) is an unusual polymerase which does not have conventional properties of a typical DNA polymerase. TdT is the template-independent DNA polymerase capable of adding DNA bases to single-stranded DNA. In nature, TdT generates variation in T and B cell receptor sequences by adding random nucleotides to ssDNA enabling our immune system to evolve and adapt. The human TdT has two splice variants, TdTS (shorter isoform) and TdTL (longer isoform). The short isoforms have nucleotide elongation activity while the longer isoform possess $3' \rightarrow 5'$ exonuclease activity for nucleotide removal. TdT is a member of the X-family of polymerases, with structural similarities to the thumb, finger, and palm domain present in all the template-dependent β polymerases. Additionally, it has lariat-like loop and "index finger" domain giving it a ring structure allowing dNTPs to diffuse into the enzyme's active site and preventing it from binding to DNA template [16]. TdT, unlike all DNA polymerases have a unique ability to use Co^{2+}, Mn^{2+}, Zn^{2+}, and Mg^{2+} for unbiased and non-templated addition of nucleotides. However, TdT prefers to incorporate dGMP and dCMP as compared to dAMP or dTMP. The TdT method is recently adapted to multiplexed enzymatic DNA synthesis where the cofactor Co^{2+} is caged inside a photosensitive DMNP–EDTA complex. Upon exposing to UV light, the DMNP–EDTA complex undergo photolysis and release the Co^{2+} ions leading to activation of Terminal deoxynucleotidyl Transferase (TdT) leading to synthesis of oligonucleotide. To regulate the elongation, the UV light source is turned off, followed by addition of excess DMNP–EDTA to quench Co^{2+} ions converting TdT to an inactive state. This cycle is repeated till the full sequence is synthesized [17]. The artificially synthesized DNA is then cleaved from the solid support and purified by HPLC. Since the reaction is carried out in aqueous mild conditions, several restrictions of chemical DNA synthesis can be avoided.

Disadvantage
- The DNA with repeat sequences, high or low GC-content, or secondary structures is not efficiently synthesized.

3.4 Summary

The isolation of genomic DNA from any cell is one of the most common and relatively simple procedures in current genera of techniques available in the field of Genetic Engineering. But the isolation of target DNA of interest from a long stretch of bulky genomic DNA is a process that has revolutionized biology research. This chapter has made an attempt to portray the available fundamental technology of obtaining genes of interest and refinement of these techniques over several decades. The success of the recombinant DNA technology started by making a genome library, which provided a huge source of gene inserts for exploring their function, structure of proteins encoded by respective genes and interaction between them which allowed them to participate in cascade of events in a cell. Another revolutionary technique that markedly improvise the field of DNA recombination was amplifying the DNA using PCR. Implementation of the PCR methodology has provided a simple, accurate, and efficient tool to obtain multiple copies of DNA fragments. PCR is not only utilized for preparation of DNA inserts but used for probing genes, sequencing DNA and screening of recombinant DNA molecules. Recent breakthrough in genetic engineering is the possibility to artificially manufacture genes which are difficult to clone or amplify from DNA substrates but simple to synthesize. DNA synthesis also allows to fabricate biological parts and circuits that do not exist naturally, as well as allows to mix and match of parts in order to develop a nonconventional biological system. The refinement of tools and rapid evolution in this field continuously provides novel platforms that are more accurate, rapid, and outperforming the previous methods. The advantages of availability of surplus methods to obtain insert DNA is improving the precision to manipulate DNA along with cutting down the cost of DNA cloning workflows.

Self Assesment

Q1. Why plant cells grinded during DNA isolation?

A1. The plant cell has a tough cell wall that must be disrupted in order to release the cell content. To disintegrate the plant tissue and to break the cells, it is usually grinded with mortar and pestle. Additionally, cryogenic grinding in presence of liquid nitrogen is more effective.

Q2. What is the purpose of artificial DNA synthesis?

A2. Chemical or enzyme-based DNA synthesis allows researchers to build DNA without reference template. Both the methods allow you to produce natural and nonnatural DNA sequences producing novel proteins. As compared to DNA amplification methods gene synthesis is a lot faster and cost effective.

References

1. Green MR, Sambrook J. Molecular cloning: a laboratory manual. 4th ed. Cold Spring Harbor Laboratory Press; 2012.
2. Baxevanis AD, Ouellette BF. Bioinformatics: a practical guide to the analysis of genes and proteins. New York: Wiley; 1998.

3. Dairawan M, Shetty PJ. The evolution of DNA extraction methods. Am J Biomed Sci Res. 2020;8:39–45.
4. Boom R, Sol CJA, Salimans MMM, Jansen CL, Wertheimvandillen PME, Vandernoordaa J. Rapid and simple method for purification of nucleic-acids. J Clin Microbiol. 1990;28:495–503.
5. Miller S, Dykes D, Polesky H. A simple salting out procedure for extracting DNA from human nucleated cells. Nucleic Acids Res. 1988;16(3):1215.
6. Tan SC, Yiap BC. DNA, RNA, and protein extraction: the past and the present. J Biomed Biotechnol. 2009;2009:574398.
7. Carpi FM, Di Pietro F, Vincenzetti S, Mignini F, Napolioni V. Human DNA extraction methods: patents and applications. Recent Pat DNA Gene Seq. 2011;5(1):1–7.
8. Head SH, Komori K, LaMere SA, Whisenant T, Nieuwerburgh FV, Salomon DR, Ordoukhanian F. Library construction for next-generation sequencing: overviews and challenges. BioTechniques. 2014;56:61–4.
9. Soares MB, Bonaldo MF, Jelene P, Su L, Lawton L, Efstratiadis A. Construction and characterization of a normalized cDNA library. Proc Natl Acad Sci. 1994;91(20):9228–32.
10. Rommens JM, Iannuzzi MC, Kerem BS, Drumm ML, Melmer G, Dean M, Rozmahel R, Cole JL, Kennedy D, Hidaka N. Identification of the cystic fibrosis gene: chromosome walking and jumping. Science. 1989;245(4922):1059–65.
11. Mardis ER. DNA sequencing technologies: 2006–2016. Nat Protoc. 2017;12:213–8.
12. Kramer MF, Coen DM. Enzymatic amplification of DNA by PCR: standard procedures and optimization. Curr Protoc Toxicol. 2000;3(1):A-3C.
13. Fakruddin M, Mannan KS, Chowdhury A, Mazumdar RM, Hossain MN, Islam S, Chowdhury MA. Nucleic acid amplification: alternative methods of polymerase chain reaction. J Pharm Bioallied Sci. 2013;5(4):245–52.
14. Salic A, Mitchison TJ. A chemical method for fast and sensitive detection of DNA synthesis in vivo. Proc Natl Acad Sci. 2008;105(7):2415–20.
15. Edward AM, Berdis AJ. Terminal deoxynucleotidyl transferase: the story of a misguided DNA polymerase. Biochim Biophys Acta. 2010;1804(5):1151–66.
16. Fowler JD, Suo Z. Biochemical, structural, and physiological characterization of terminal deoxynucleotidyl transferase. Chem Rev. 2006;106(6):2092–110.
17. Lee H, Wiegand DJ, Griswold K, Punthambaker S, Chun H, Kohman RE, Church GM. Photon-directed multiplexed enzymatic DNA synthesis for molecular digital data storage. Nat Commun. 2020;16(1):1–9.

Chapter 4
DNA Cutters in Recombinant DNA Technology

Sangeeta Sathaye, Aruna Sivaram, and Nayana Patil

4.1 Introduction

Restriction enzymes are also referred to as molecular scissors or restriction endonucleases. Restriction enzymes recognize specific sequences of nucleotides in a DNA and cut the DNA into fragments at these sites or more randomly. These specific sequences of nucleotides recognized by the enzymes in the DNA are known as restriction sites. These sites are specific for specific restriction endonucleases. The enzymes cut both the strands of DNA by hydrolysis of the phosphodiester bond. Before these enzymes were discovered there was no known method for cutting DNA at discrete specific sites. The usefulness of restriction endonucleases in DNA manipulation was quickly recognized. The restriction enzymes are present in many bacteria and Archaea and to date several have been successfully isolated and are being used in DNA manipulation.

In the early twentieth century, it was shown that some viruses called bacteriophages infect bacteria in a host specific manner. These viruses inject their DNA into the bacteria, multiply using the host cell machinery and then leave the host cell causing the death of the host. In nature, the restriction modification system evolved as a defensive mechanism in archaea and bacteria to fight viral infections. Using this system, the bacteria can monitor incoming DNA and destroy the virus if the DNA is recognized as foreign. The detailed mechanism of restriction modification was explained in the 1960 [1].

S. Sathaye
Modern College of Arts Science and Commerce, Shivajinagar, Pune, Maharashtra, India

A. Sivaram · N. Patil (✉)
School of Bioengineering Sciences & Research, MIT ADT University,
Pune, Maharashtra, India
e-mail: aruna.sivaram@mituniversity.edu.in; nayana.patil@mituniversity.edu.in

N. Patil, A. Sivaram, *A Complete Guide to Gene Cloning: From Basic to Advanced*,
Techniques in Life Science and Biomedicine for the Non-Expert,
https://doi.org/10.1007/978-3-030-96851-9_4

Restriction modification system involves the secretion of two special enzymes by the bacteria—an endonuclease and a methylase. The endonucleases recognize a specific sequence in the viral genome and cleave it to protect the host from the viral attack. The bacteria can protect themselves from these endonucleases by modifying their own DNA by a special process of methylation of specific adenine or cytosine bases catalyzed out by the second enzyme of the host modification system, namely the methylases. The endonucleases were then called the restriction enzymes.

Information on the mechanism of restriction and modification suggested that the bacterial systems not only limit the rate of viral infection but also could allow the exchange of genetic information between related species of bacteria with the same restriction and modification systems.

In 1970, Werner Arber, Daniel Nathans, and Hamilton O. Smith were awarded the Nobel Prize for Physiology and for Medicine in 1978 for their discovery and characterization of the first restriction enzyme HindII. After the discovery of HindII, many other restriction endonucleases were isolated and characterized. This eventually led to the development of DNA manipulation and recombinant DNA technology.

Restriction modification systems are an extremely diverse group of enzymes and widespread among bacteria and Archaea. Several restriction enzymes have been reported to date [2]. Among the four different types of restriction and modification systems that have been recognized only Type II is widely used in gene manipulation.

The nomenclature of restriction enzymes was suggested by Smith and Nathans in 1973 [3]. According to their proposal, the names of the enzymes begin with a three-letter acronym written in Italics. The first letter stands for the bacterial genus from which the enzyme has been isolated and the next two letters stand for the bacterial species. The name can also be followed by extra letters or numbers to indicate the serotype or strain. This is followed by a space and then a number written in Roman numerals to indicate the chronology of identification.

For example, *in the enzyme BamHI*

Bam stands *for Bacillus amyloliquefaciens* the bacterium it was isolated from

"H" represents strain H and "I" stands for the first enzyme isolated from this species

4.2 Classification Based on the Features of Enzymes

Restriction enzymes vary broadly in their structure, unique site for cleavage and on the co-factor required for their action. Based on these criteria, these enzymes are classified into 4 types (Table 4.1). Besides being of great importance in DNA manipulation, restriction enzymes have proven to be excellent systems to study highly specific protein–nucleic acid interactions, to investigate structure–function relationships and to understand how these enzymes evolved.

Table 4.1 Different classes of restriction endonucleases

Class	Characteristics	Examples
I	• Different subunits for recognition, cleavage and modification • Recognize and modify same sequence • Cleave bases at a distance away from the recognition site about 1000 bp • Require ATP for activity	*Eco*BI
II	• Two different enzymes-endonuclease and methylase • Both recognize the same site • Most recognize a site which is symmetrical • Can work independent of each other • No ATP required for activity	*Bam*HI *Eco*RI *Hin*DIII
III	• Two subunits—One each for recognition and modification and for cleavage • The recognized sequence is methylated • Cleaves 24–26 bp away	*Hinf*III
IV	• Two different enzymes • Recognition sequence is asymmetric often methylated • Cleavage occurs on one side of recognition sequence up to 20–30 bp away and is often methylated	*Mcr*BC

4.2.1 Type I Enzymes

The Type I enzymes are composed of five subunits. The active enzyme consists of two restriction subunits (2R), two subunits that can modify the DNA by methylation (2 M) and one recognition subunit (S). They possess both the methylase and the endonuclease activity. These enzymes show a requirement for ATP, magnesium ions as well as S-adenosyl methionine for their activity.

The Type I enzymes recognize two sets of sequences separated by a gap. The S subunit consists of a pair of target recognition sequences that each recognizing one half sequence [4].

Enzyme *Eco*KI	IA	AAC N6 GTGC	N= any base
		AAC NNNNNN GTGC	

Recombination between target recognition sequences can generate new sequence specificities in vivo and is the cause of Type I R–M system diversification. The Type I enzymes are further divided into five groups based on their amino acid sequence, antibody reactivity and complementation tests. When nucleic acid treated with these enzymes are electrophoresed, distinct restriction fragments or different band patterns are not formed. This class of enzymes is of less practical value.

4.2.2 Type II Enzymes

The classification of restriction enzymes to Type II is based on the property of the enzyme to cleave specifically within or close to the recognition site. This process does not require ATP. The recognition sequences of Type II enzymes are mentioned in Table 4.2.

A typical Type II enzyme recognizes a sequence which is 4–8 bp long, palindromic, cuts within the sequence or adjacent to the sequence to generate 5′-phosphate and 3′-OH ends and requires the presence of Mg^{2+} ions.

The catalytic activity generates either blunt ends or sticky ends with overhangs.

Blunt ends are formed when the restriction enzymes cut the double-stranded DNA in a vertical line to generate 3 prime free hydroxyl and 5 prime phosphate groups.

Example: *PvuII* [CAG * CTG]

Sticky ends are generated as small single strands which can join with complementary bases of another DNA or self-ligate. The overhangs can be at the 5′-phosphate or 3′-hydroxyl ends.

Example: *PstI produce* 3′-OH overhangs [*CTGCA * G*]

EcoRI generate 5′-phosphate overhangs [G * AATTC]

Subclasses of Type II enzymes: Some Type II endonucleases do not show the basic property and have been further classified into subtypes [5] as given below.

Type IIS: **This class of Type II enzymes recognize sequences which are not symmetric and cut these sequences at a definite distance.**

Example – *Fok*I

Recognition site

GGATG N9 * NNNN

CCTAC N9 NNNN *

Table 4.2 Recognition sequences of Type II enzymes

Name of the enzyme	Number of nucleotides in recognition sequence	Sequence and splice site
HaeIII	4	GG* CC
AluI	4	AG*CT
BamHI	6	G*GATCC
ClaI	6	AT*CGAT
SbfI	8	CCTGCA*GG

*- site of cut/splice site

- Recognition of 5′-GGATG-3′ site by the enzyme activates its cleavage domain. As a result, it cleaves at a position of 9 nucleotides upstream of the recognition sequence in the first strand and 13 nucleotides downstream of it in the second strand.
- *Type IIE*: This subtype recognizes and interacts with their recognition sequence in two facets—a target for cleavage and an allosteric effector. Example *Nae*I.

NaeI Recognition site GCG * CGC

CGC * GCG

- *Type IIF*: These are similar to type IIE enzymes and require interaction with two parts of their recognition sequence. They differ from type IIE enzymes as they cut both sequences in a concerted reaction.

Example *NgoMIV*

Recognition site G * CC GGC

CGG CC*G

- *Type IIT*: These restriction endonucleases are made up of two different subunits—one with a restriction and the other with a modification activity.

Example *Bpu101*

Restriction site

CGC*TANAGCCG

G NT*

- *Type IIB*: This type of restriction endonucleases cut DNA on both sides of the recognition sequence which may be symmetric or asymmetric.

Example *BcgI*

Recognition site
NN*N(10)CGAN(6)TGCN(10)NN*
*NN N(10)GCTN(6)ACGN(10)NN

- *Type IIG*: Enzymes belonging to this subtype have restriction and modification activity present in a single polypeptide chain.

- *Type IIM*: These restriction endonucleases recognize only methylated DNA.

Example *DpnI*

Recognition site G ^{m}A* TC

CT * ^{m}AG

4.2.3 Type III and Type IV

Type III enzymes require the ATP for their activity. The recognition sequence and cleavage site of these enzymes is 20–30 bases apart. The recognition sites are inversely oriented and unmethylated. Example: EcoP15.

Type IV enzymes recognize modified, typically methylated DNA. They are also known as methylation dependent restriction enzymes (MRDE). Their cleavage is non-specific. Example McrBC and Mrr systems of *E. coli*. McrBC isolated from Escherichia coli K-12 is an endonuclease which cuts methylated cytosine DNA region on one or both strands. It requires only methylated DNA and has a unique ring-like structure.

These enzymes are less characterized and they are not frequently used in recombinant DNA technology.

4.3 Classification Based on Recognition Site and Cleavage Specificity

Some enzymes in the family of restriction enzymes are similar in their choice of recognition sites or generate similar end products and based on recognition site and cleavage specificity they are grouped as isoschizomers, neoschizomers, and isocaudomers.

4.3.1 Isoschizomers

These are a pair of enzymes isolated from different organisms but having the same sequence recognition and cleavage site, and may require altered conditions for enzymatic activity. By convention, the prototype is the first enzyme and the recognition sequence discovered. The enzymes subsequently discovered are the isoschizomers. Example: *Sph*I and *Bbu*I have a common cleavage site CGTAC/G. *Sph*I was isolated from *Streptomyces phaeochromogenes* while *Bbu*I from *Bacillus* spp.

Another pair of restriction enzymes *HpaII* and *MspI*, recognize the sequence 5′-CCGG-3′, *MspI* can recognize it only when the second cytosine is methylated, while *HpaII* cannot.

4.3.2 Neoschizomers

Restriction enzymes that have the same recognition sequence but cleave differently are neoschizomers.

Example: the enzymes SmaI and XmaI both recognise the sequence CCCGGG but the cleavage as follows:

SmaI **CCC *GGG**

XmaI **C * CCGGG**

Both generate different types of ends (blunt ends for SmaI and 5′ protruding ends for XmaI).

4.3.3 Isocaudomers

These are restriction enzymes that have different recognition sequences, but upon cleavage of DNA, generate identical end products. These sequences can be joined to one another and result in a new asymmetrical sequence that cannot be cut by a restriction enzyme. Such pairs can be of use to alter or remove restriction sites.

Example: *BamH I*	*MboI*
G *GATC *C	N * GATC * N
C *CTAG *G	N * GATC *N

Box 4.1
In silico tools have simplified the use of type II restriction enzymes for molecular biology and cloning. Information regarding the enzymes, their cleavage sites, choice of appropriate enzymes, etc. can be identified using these tools. A very exhaustive collection of information about these enzymes including their specific sites for recognition, effect of methylation of DNA modification have been given in a database called REBASE. Another tool which is very commonly used by the researchers and scientists is the NEBcutter. This is a web-based, freely available, user-friendly tool using which one can analyze if a DNA sequence contains specific cleavage sites for restriction enzymes. Other web-based tools for analysis of restriction sites include dCAPS Finder, BlastDigester, SNP2CAPS, SNP Cutter, etc. In these web-based tools, only small datasets can be uploaded. Certain tools like Customizable in silico Sequence Evaluation for Restriction Sites (CisSERS) enable high-throughput analysis of multiple samples. These restriction enzyme analysis tools have become an indispensable part of any cloning or molecular biology work where these enzymes are involved.

4.4 Novel Restriction Endonucleases

A few restriction enzymes have also been reported from non-bacterial and non-archeal origins. Following are some examples:

SacC1 was isolated from yeast (*Saccharomyces cerevisiae)*, a eukaryote and recognizes the palindromic sequence 5′CTCGAC3′. This enzyme generates a staggered cut five bases away from the recognition site on one strand of DNA and seven bases away on the complementary strand. This enzyme shows similar characteristics with *Psp124B1* from *Pseudomonas* species [6].

HsaI a restriction enzyme was reported to be human in origin and was isolated from human embryos. It is an isoschizomer of *EcoRI* with a distinctive elution profile from *EcoRI* [7].

4.5 Artificial Restriction Enzymes

Foreign DNA can be introduced into the acceptor DNA by the use of restriction enzymes and their specific cutting sites. This process has been used as a gene editing tool using over 250 naturally occurring restriction enzymes that are commercially available. So far over 3600 restriction endonucleases have been isolated and characterized from several bacterial and archaea species.

The use of restriction endonucleases can be a limitation, mainly due to a short (4–8 basepairs) recognition site which can appear multiple times in a long stretch of DNA. This has been addressed recently by artificial restriction enzymes (AREs) that have recognition sites with longer sequences thereby increasing specificity, resulting in very few cleavage sites and longer sticky ends. These enzymes are engineered by fusing DNA binding domain with a nuclease domain. These AREs can target long stretches of DNA (upto 36 bps) and bind desired areas of the DNA.

Restriction endonuclease has led to the development of exciting gene editing tools like. The popular examples are Zinc finger nucleases, the CRISPR/Cas9 a bacterial immune system and TALENs (transcription activator-like effector nucleases).

In the year 2010, Murtola, Wenska, and Strömberg reported PNA-based systems (PNAzymes) that act as site and sequence specific RNAses. These enzymes carry a Cu (II) -2-9- dimethylphenanthroline group and cut no base paired regions in the RNA (RNA bulges) and could be the basis of future RNA specific restriction enzymes [8].

In 2017, Enghiad and Zhao of Illinois University demonstrated a new method of creating AREs using an Argonaute protein (PfAgo) isolated from a species of archea *Pyrococcus furiosus*. PfAgo led by a DNA guide is able to recognize long sequence cleavage sites and thus increasing specificity of region to be cut. PfAgo also creates longer sticky ends [9].

4.6 Applications of Restriction Enzymes in Recombinant DNA Technology

The property of restriction endonucleases to cut DNA at specific recognition sites has enabled the use of these enzymes in molecular biology techniques. Some of the major applications are as follows:

4.6.1 Cloning

This is one of the most useful applications of restriction enzymes used in recombinant DNA technology for generating of recombinant DNA molecules. The technique involves cutting of the donor DNA and the vector DNA by the same restriction enzyme resulting in compatible ends which could be used to join the vector and donor DNA by an enzyme called DNA ligase. The recombinant DNA generated can then be maintained in the host for replication or used for transformation. A large number of vectors can be created using cloning methods. However, this method cannot be used if the acceptor DNA does not have the same active site as that of the donor.

4.6.2 *Restriction Mapping*

Restriction mapping is a technique used to create a map of the fragments generated by treating the DNA of interest with a set of restriction enzymes. The map is created by running the treated DNA on an agarose gel causing the fragments of various sizes to form bands on the gel. This gives information of the relative positions of the restriction sites on the DNA of interest. Thus restriction endonucleases can be used to obtain structural information of the DNA fragment or verify the identity of the DNA fragment. Restriction maps of a DNA segment are unique for the set of restriction enzymes used. The use of restriction mapping is limited if the DNA has few restriction sites in its sequence.

4.6.3 *Restriction Fragment Length Polymorphism (RFLP)*

This application also makes use of restriction maps created by an individual or a set of restriction enzymes on various DNA segments. The maps obtained can be used to check for uniformity of the DNA segments used or detect polymorphisms. This is done using probes. The digested fragments are separated on the gel based on size and are transferred onto a membrane. These fragments are then tagged with a radioactive or fluorescent probe targeting specific regions near or bracketed by restriction enzyme sites. Each DNA under consideration (generally belonging to an individual) has a unique pattern called the "biological barcode." A polymorphism occurs when this unique pattern varies between individuals.

The technique has been used for profiling DNA of individuals and can then be applied for gene mapping, localization of genes for genetic disorders, determination of risk for disease, and in forensic science for paternity testing. It has also been used as a testing tool for detecting somaclonal variation in regenerated plants. This is not a high throughput system and will have limited use if several individuals are to be screened for polymorphisms.

4.7 Summary

The discovery of restriction endonucleases and their mechanisms demonstrated that DNA from different sources could be cut by the same restriction enzyme and joined together by ligase enzymes. This was possible because of the complementarity of the cut ends of the different DNA molecules. This complementarity was a very valuable tool in the hands of the molecular biologists for creating recombinant DNA molecules. New techniques such as recombinant DNA technology, DNA diagnostics, DNA forensics, to name a few, have evolved and have found immense applications.

Self Assesment

Q1. Given below is a DNA sequence. If this were to be cut with a restriction enzyme BamHI how many fragments would be generated? BamHI recognizes GGATCC sequence and generates sticky ends

5′ TTGAAAAGGATCCGTAATGTGTCCTGATCACGCTCCACG 3′
3′ AACTTTTCCTAGGCATTACACAGGACTAGTGCGAGGTGC 5′

A1. The two fragments are

5′ TTGAAAAG3′ 5′ GATCCGTAATGTGTCCTGATCACGCTCCACG 3′
3′ AACTTTTCCTAG 5′ 3′GCATTACACAGGACTAGTGCGAGGTGC 5′

Q2. Given below is a linear DNA having restriction sites for *Eco*RI and *Bam*HI. The DNA is cut single and then with both enzymes simultaneously and produces the following bands when run on a gel. Construct a restriction map for the given linear DNA.

Total length 18 kb
*Eco*RI: 3 fragments: 5 kb, 9 kb, 4 kb
*Bam*HI: 2 fragments: 7 kb and 11 kb
Both together: 4 fragments: 5 kb, 2 kb, 7 kb, and 4 kb

A2. If *Eco*RI generates three fragments means there are two *Eco*RI sites on this DNA. *Bam*HI generates only two bands, so there is a single site for this enzyme. The digest with both these generates fragments smaller than 11 kb, so one of the *Eco*RI site is a part of this DNA.

Possible solution shows the presence of a *Bam*HI site between two *Eco*RI sites

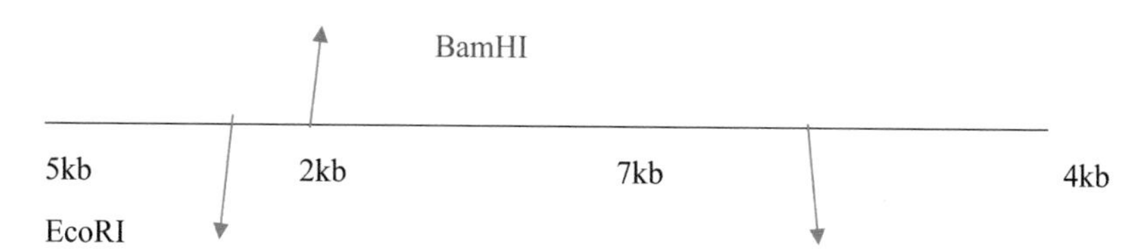

References

1. Arber W. Host-controlled modification of bacteriophage. Ann Rev Microbiol. 1965;19(1):365–78.
2. Pingoud A, Fuxreiter M, Pingoud V, et al. Type II restriction endonucleases: structure and mechanism. Cell Mol Life Sci. 2005;62:685. https://doi.org/10.1007/s00018-004-4513-1.
3. Smith HO, Nathans D. Letter: a suggested nomenclature for bacterial host modification and restriction systems and their enzymes. ACS Synth Biol. 2017;6(5):752–7.
4. Roberts RJ, Vincze T, Posfai J, Macelis D. REBASE—a database for DNA restriction and modification: enzymes, genes and genomes. Nucleic Acids Res. 2010;38:D234–6.
5. Kan NC, Lautenberger JA, Edgell MH, Hutchison CA III. The nucleotide sequence recognized by the Escherichia coli K12 restriction and modification enzymes. J Mol Biol. 1979;130(2):191–209.

6. Shikara M. Identification of a restriction endonuclease (SacC1) from Saccharomyces cerevisiae. J Yeast Fung Res. 2010;1(7):127–35.
7. Lao WD, Chen SY. HsaI: a restriction enzyme from human being. Sci Sin Ser B Chem Biol Agric Med Earth Sci. 1986;29(9):947–53.
8. Murtola M, Wenska M, Stromberg R. PNAzymes that are artificial RNA restriction enzymes. J Am Chem Soc. 2010;132(26):8984–90.
9. Enghiad B, Zhao H. Programmable DNA-guided artificial restriction enzymes. ACS Synth Biol. 2017;6(5):752–7.

Chapter 5
Overview of Gene Cloning Strategies

Neeraj Maheshwari, Praveen Kumar, Aruna Sivaram, and Nayana Patil

5.1 Introduction

The production of first rDNA molecules using restriction enzymes was carried out in early 1970s, reinforcing focus on molecular genetics. Thereafter, RDT is one of the fields that has extensively ramped up in its utilization and complexity, yielding increasingly potent methods for DNA manipulation. It has multidisciplinary applications in agriculture, hormones, vaccines, therapeutic agents, antimicrobial peptides, and recombinant diagnostic probes [1]. Discovery of bacterial enzymes that can cleave DNA molecules at specific positions led to a significant development in molecular cloning techniques. Molecular cloning is the most crucial and popular set of techniques to bring together DNA molecules forming recombinants and hybrids DNA constructs capable of performing a plethora of functions. By definition, molecular cloning is a process in which recombinant DNA molecules of interest are assembled in vitro and replicated into a host organism. This process constitutes two elements: isolation and amplification of a specific DNA fragment to be replicated and a vector for propagation.

By means of restriction enzymes the DNA fragments of interest are isolated from the source, copied, and amplified using PCR. Upon isolation, clones can be used to generate numerous copies of the DNA for analysis, and/or to express the proteins

N. Maheshwari · A. Sivaram · N. Patil (✉)
School of Bioengineering Sciences & Research, MIT ADT University,
Pune, Maharashtra, India
e-mail: neeraj.maheshwari@mituniversity.edu.in; aruna.sivaram@mituniversity.edu.in;
nayana.patil@mituniversity.edu.in

P. Kumar
Kashiv Biosciences, Ahmedabad, Gujarat, India
e-mail: praveen.kumar@kashivbio.com

N. Patil, A. Sivaram, *A Complete Guide to Gene Cloning: From Basic to Advanced*,
Techniques in Life Science and Biomedicine for the Non-Expert,
https://doi.org/10.1007/978-3-030-96851-9_5

for the study or employment of its function. Moreover, site-directed mutations of the clones allow them to detect the function of target protein, by altering the quantity and quality of protein.

However, the choice of restriction enzymes is more critical for designing and cloning thereby increasing the efficiency by generating complementary "sticky ends" [2]. In sticky ends the DNA has a single stranded overhang on either 3′ or 5′ ends. The sticky ends must be converted into blunt ends by either by removing the overhangs or by inserting complementary base pairing.

In RDT, a gene of interest is obtained by splicing it from source, copying, or assembling it using oligonucleotides and inserting it into a suitable vector. The DNA becomes an integral part of the new vector by phosphodiester bonds and is replicated by the host. The vectors can be prokaryotic (plasmid, bacteriophages, cosmid vectors) or eukaryotic (yeast or mammalian artificial chromosomes). Generally, plasmids (replicates independently of chromosomal DNA) are introduced as vectors into the bacterial host. Important feature in plasmid is the presence of a short segment of DNA which contains multiple restriction sites called multiple cloning sites, also known as a polylinker.

Basically, molecular cloning includes four fundamental steps:

1. Isolation of insert or target DNA fragments
2. Ligating the insert into suitable vector plasmids
3. Transformation of recombinant plasmids into host for multiplication
4. Identify the correct host cell carrying the recombinant molecule

Nowadays, different molecular cloning strategies have been developed for different purposes. It is important to note that for a single cloning project, a combination of several methods may actually yield the best results. The choice of most appropriate cloning strategy would depend on various factors like the efficiency of cloning, availability of infrastructure and reagents, the efficiency of each method and the time available with the researcher. In this chapter, we will be walking you through different cloning strategies in detail.

5.2 PCR Cloning

PCR cloning is a versatile technique and has been widely used for biological engineering. It allows DNA fragments to be inserted into the backbone of the vector even when it is in minimal quantity [3]. In this method, a PCR generated DNA fragment is directly ligated into the vector. One of the most commonly used and the simplest form of PCR cloning techniques is the TA cloning. The method requires designing of appropriate primers and optimization of the PCR conditions. During amplification of a template, Taq polymerase favors addition of an adenosine molecule at 3′ ends. These A sites can be directly ligated with T-tailed vectors, hence also known as TA cloning. This method utilizes the advantage of hybridizing adenine and thymine on different fragments in presence of ligase. The TA cloning method can be easily modified and is especially useful for restriction digestion free insertion of fragments in vector DNA. It is a convenient and less laborious method, and can be performed even with a limited quantity of the starting material [4]. This method also helps in avoiding the restriction digestion of the insert and the vector. However, one of the major disadvantages is that this method cannot be used for directional cloning. The DNA fragment used for cloning should be less than 5 kb. As this method relies on Taq DNA polymerase which does not have a proofreading activity, the error rate may also be high. In order to reduce the errors, high-fidelity enzymes can be used for amplification. These enzymes do not produce the Adenosine overhangs. These blunt ended DNA fragments are ligated with linearized vectors (which are also blunt ended). This may result in lower cloning efficiency. Several commercially available vectors are available which have improved the efficiency of cloning of blunt ended and sticky ended DNA fragments. Figure 5.1 gives the details of PCR cloning.

Long-term storage and presence of endonucleases may cause degradation of nucleotide overhangs of PCR products, thereby reducing the efficiency. Hence fresh and purified PCR products in TA cloning are recommended. Using terminal transferase, the T vector is prepared allowing T and A annealing and ligation. The unidirectional in TA cloning can be achieved by manipulating the phosphorylation status of the DNA molecules ensuring cloning of insert in right the orientation.

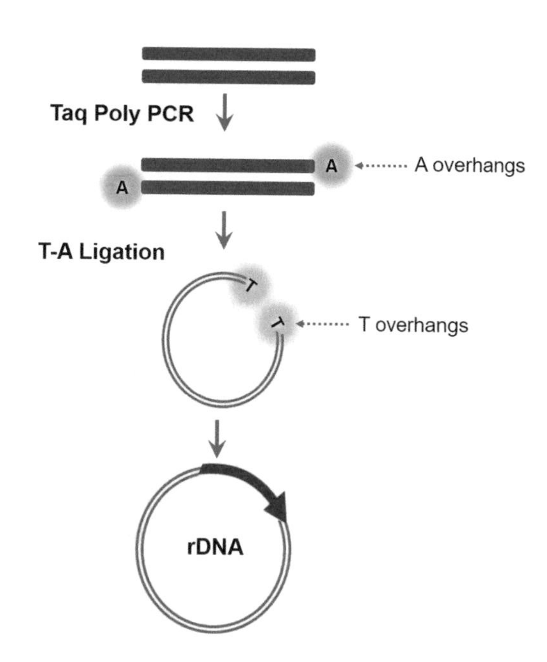

Fig. 5.1 TA Cloning method: Illustration of TA-mediated cloning. The PCR catalyzed by Taq polymerase add an A overhang to the products. Vectors with T overhang match with the PCR product and are ligated to obtain recombinant plasmid

5.3 Restriction and Ligation-Based Cloning

Special bacterial enzymes that cut DNA at specific sites (known as restriction sites) are called restriction enzymes. Depending on the ends generated after digestion, the restriction enzyme is termed as sticky end cutters or blunt end cutters. As the name suggests a blunt end cutter generates blunt ends while sticky end cutters create 3′ or 5′ overhang. Endonuclease restriction enzymes make a cut within the DNA while exonuclease restriction enzymes chop the DNA from an end. The restriction enzymes have been covered in detail in Chap. 5.

Restriction enzyme-based cloning is one of the most popular cloning strategies, with very versatile applications. In this technique, both the DNA fragment to be cloned and the vector are treated with appropriate restriction enzymes. After digestion, the insert and the vector are ligated with DNA ligase, which will result in a circular vector which contains the insert. This cloned vector can be maintained in different biological systems, like the bacteria *E. coli*.

Restriction enzymes like the *Eco*RI recognize conserved short inverted repeats (also known as palindromes) generating reproducible set of fragments called as restriction fragments. Upon digestion with the same enzyme, the same complementary tails are generated in insert and vector DNA. At room temperature, these compatible ends readily base-pair with each other. This base pairing of sticky ends permits DNA from widely differing species to be ligated, forming chimeric molecules.

Other restriction enzymes, such as *Alu*I (5′-AG/CT-3′) and *Sma*I (5′-CCC/GGG-3′), generated fragments with blunt ends. In these ends, all the complementary nucleotides are base paired. Blunt end cloning involves ligation of both strands of DNA into the vector with no overhanging bases at the termini. The phosphodiester bond of nearby 5′ phosphate and 3′ hydroxyl groups of cut DNA in presence of T4 ligase (isolated from phage T4) is sufficient to produce clones. T4 DNA ligase work in sequential steps of (a) enzyme adenylation: addition of adenosine monophosphate (AMP) molecule from ATP or NAD+ into ligase, (b) transfer of AMP: the AMP molecule is transferred at the 5′ phosphate at the site of nick, (c) nucleophilic attack of 3′-OH on 5′ phosphate to form phosphodiester bond. Blunt end cloning preparation is easy because it avoids enzymatic digestion and subsequent purification needed in sticky end cloning.

T4 DNA ligase is a ligating enzyme that can join any two cohesive or blunt DNA ends by forming phosphodiester bonds between adjacent nucleotides (Fig. 5.2). This enzyme will not join single stranded DNA. Blunt end ligation, on the other hand, demands a higher DNA concentration than sticky end ligation [3].

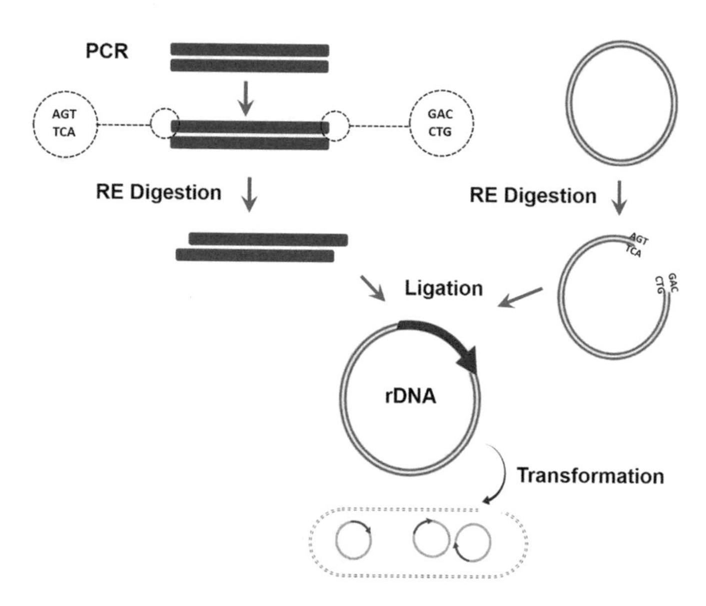

Fig. 5.2 The insert DNA is PCR amplified using primers with restriction site linkers. The vector is digested with the same enzyme to generate compatible ends. Next the insert and vector are ligated together using the matching ends and introduced in host cells

5.4 Ligation-Independent Cloning

The success of restriction digestion and ligation cloning method depends on the presence of appropriate restriction site selection and efficient ligation between the DNA molecules [5]. In case the DNA molecule does not have specific restriction enzymes the method would fail to produce recombinant plasmids. Therefore, to overcome this lacuna, Ligation-Independent Cloning (LIC) was developed which is highly efficient, simple, and faster. In this technique short sequences of DNA (12 nucleotides long) are added to the clone fragments that are homologous to the vector. Complementary and compatible ends are generated by $3' \rightarrow 5'$ exonuclease activity of T4 DNA polymerase. The resulting molecules are mixed together to form non covalent association and annealed. T4 DNA Polymerase's polymerization and exonuclease activity are balanced by exonuclease processing to the first complementary C residue by addition of dGTPs. Thus the vector has four nicks on each strand which are repaired by host (*E. coli*) during transformation (Fig. 5.3). In this technique the resultant recombinant plasmid is "scarless" as it does not contain any unwanted sequences or new restriction enzyme sites [3]. Ligation-independent cloning. LIC includes polymerase incomplete primer extension (PIPE) cloning which is a two-step process. Hybrid vectors are formed when complementary strands anneal overlapping sequences which are introduced at the ends of incomplete extensions. These hybrids are directly transformed into the host without any enzymatic manipulations [6], sequence and ligation-independent cloning (SLIC)

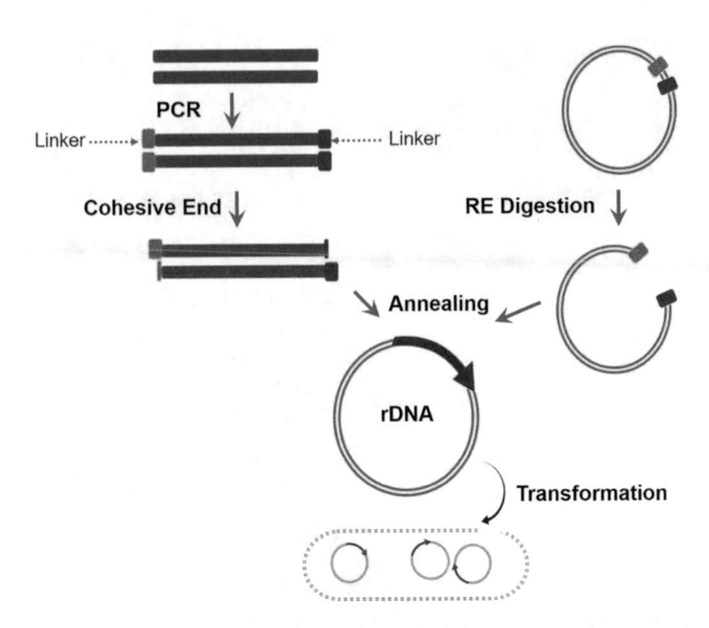

Fig. 5.3 Ligation-Independent Cloning (LIC) Workflow: First the insert DNA is PCR amplified using LIC primers. The PCR product and vector are treated with 3′ to 5′ exonuclease to create cohesive overhangs. The nicks in the complementary overlap are sealed by annealing to obtain rDNA which can be multiplied by transforming in competent host cells

uses in vitro homologous recombination and single strand annealing to assemble multiple fragments of DNA in a single reaction [7], and overlap extension cloning (OEC) employs chimeric primers to the insert creating overlapping regions with the vector. By mixing the insert and vector, hybrids are generated. The inserts are then extended using Phusion DNA polymerase, using vectors as template until polymerase reaches 5′ end forming a new plasmid. The parent plasmid is degraded by DpnI restriction enzymes leaving behind newly synthesized plasmid which can now be transformed into the host [8].

LIC is based on the exonuclease activity of T4 polymerase generating complementary overhangs without dependency on presence of specific restriction sites [9]. The extended length of compatible ends holds the insert and vector together, the nick between the nick is sealed by the DNA repair machinery of the host in which the rDNA is transformed. As background recombinants are not formed, thereby reducing the need to screen for recombinants [10].

Advantages

- It is highly efficient, allowing direct transformation into the host without any in vitro ligation.
- It does not cleave the insert sequence using restriction enzymes.
- It can be used to clone a library of unknown sequences.

Disadvantages

- Every fragment needs to be sequenced when used in different vectors.
- Cloned fragments cannot be recombined as in other techniques (Gateway).

5.5 Recombinational Cloning

Traditional methods of cloning allow handling only a few genes of interest at a time [11]. Recombinational cloning gives flexibility to insert multiple genes using site-specific recombination into multiple expressions and cloning systems simultaneously. These enzymes are capable of swapping and shifting single DNA sequences into multiple expression systems or multiple DNA inserts into a single expression vector at the same time, with widely available open reading frame (ORF) collections [12].

Gateway cloning system (Invitrogen) and Creator system (BD Clontech) is the most widely used system in this category [12]. It is based on a site-specific recombination system used by phage to integrate DNA in bacterial host cells. This technique relies on two proprietary enzyme mixes BP and LR clonase to transfer DNA fragments across recombination sites. Initially appropriate sequence of interest is cloned into a holding vector ("Entry" for Gateway) using traditional cloning methods. Once the new clone is created, it can be easily transferred to a variety of "destination" or "acceptor" vectors that include sequences which can be identified by recombinase (Fig. 5.4). This in vitro version of integration and excision reactions are made directional by developing att1 and att 2 recombinational sites.

Advantages

- Robust and high cloning efficiency.
- It allows to maintain the desired reading frame and orientation.

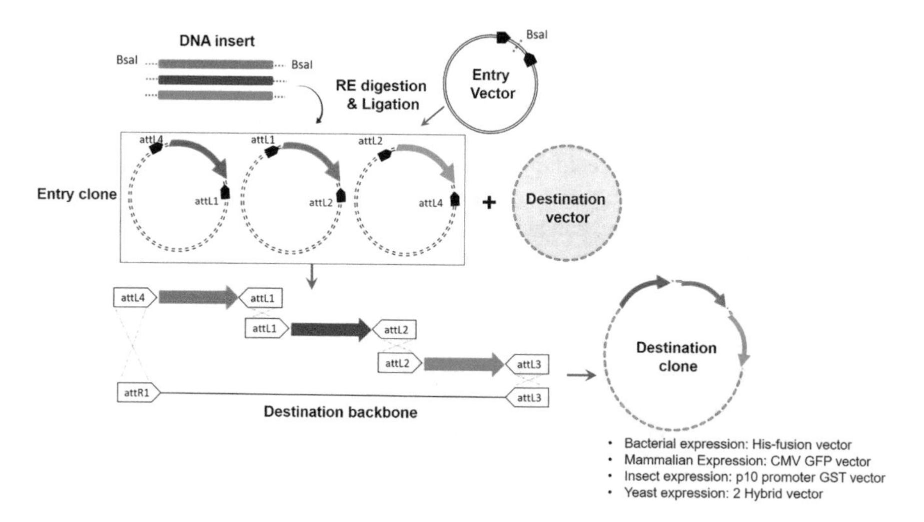

Fig. 5.4 Recombinational cloning workflow: In the first step Insert DNA sequences are cloned into entry vectors by traditional cloning methods or LIC. Using the flanking BsaI restriction sites, multiple fragments can be sub cloned as desired. The site-specific recombination between the flanking att sequences allows rearrangement and transfer into various destination vectors to create the final expression vector

- Single entry clone genes can be easily subcloned into a variety of destination vectors [13].

Disadvantage
- It is difficult to switch to another recombination system (lack convenient restriction endonuclease sites and start and stop codons are removed).
- Recombination enzymes are very expensive where the vector sets are defined by the supplier and often require proprietary enzymes to be used [14].

5.6 Mating-Assisted Genetically Integrated Cloning (MAGIC)

A highly engineered process of in vivo cloning "Mating-Assisted Genetically Integrated Cloning" or MAGIC was developed by Li and Elledge enabling rapid assembly of recombinant DNA molecules. During bacterial mating, it uses site-specific DNA cleavage and homologous recombination. The transfer of DNA fragments is aided by specifically made "donor" and "recipient" vectors (e.g., insert with specific recipient expression vectors and donors). The recombination events do not need any DNA preparation or in vitro manipulations. Recombination events are genetically chosen which results in the efficient positioning of the target gene under the control of new regulatory elements [7]. As this new method involves mixing of bacterial strains, it brings about a high-throughput recombinant DNA production that is seamless, saves time and effort.

Bacterial mating can efficiently transfer large DNA fragments (>100 kb). The donor strain must have conditional origin of replication and is nonfunctional in the recipient strain Two genetic systems initiate homologous recombination when the two strains are in the same cell. The first is by I-SceI site-specific endonuclease activity over donor and the recipient. The second is the lambda recombinase system which brings about homologous recombination between 50 bp. Combining all these features allows efficient and enhanced transfer of fragments from the donor vector onto the recipient [15].

Advantages
- The technique permits bacteria to mix inserts and vectors, allowing constructs to be generated without restriction enzyme digestion, gel purification, or the use of recombination-promoting enzymes [15].
- MAGIC can generate multiple constructs at the same time by using different recipient strains.

Disadvantage
- During the transfer of genes, there may be gain or loss of some nucleotides.

5.7 Summary

In this chapter, we provided an outline of various molecular cloning techniques and described fundamental working mechanisms for each. It also provides an overview of the progress made in development of specialized cloning techniques to improve efficiency by exploiting the properties of enzymes. Molecular cloning has evolved from cloning a few sequences of single DNA fragments to assembly of multiple fragments onto a single stretch of DNA. Additionally, these methods have been advantageous for making processes simple, seamless, and ability to bring about high-throughput production of recombinant DNA in a short stretch of time. In future, these technologies will emerge into processes that are able to insert or build sequences adjacent to each other to synthesize large DNA molecules. It will provide molecular biologist tools to explore, alter, and harness DNA, further broadening the horizon of science.

Self Assesment

Q1. What are the different types of DNA cloning methods?

A1. There is the classical traditional cloning method, PCR-based cloning and restriction enzyme-based cloning along with advanced methods such as recombination-based cloning and Mating-Assisted Genetically Integrated Cloning. Each method is based on the principle of inserting target DNA into a vector, but uses different mechanisms and approaches to achieve the insertion. The advances in the method make it affordable, efficient, and time saver.

Q2. Which of the DNA cloning methods are available commercially?

A2. Various types of DNA cloning kits are available based on Traditional and advanced DNA cloning methods. Few commercial products are TOPO TA cloning kit, In-Fusion, Gibson assembly GeneArt, Gateway Echo cloning, and Creator.

References

1. Khan S, Ullah MW, Siddique R, Nabi G, Manan S, Yousaf M, Hou H. Role of recombinant DNA technology to improve life. Int J Genom. 2016;2016:2405954.
2. Lessard JC. Molecular cloning. Methods Enzymol. 2013;529:85–98.
3. Bertero A, Brown S, Vallier L. Methods of cloning. In: Basic science methods for clinical researchers. Academic Press; 2017. p. 19–39.
4. Zhou MY, Gomez-Sanchez CE. Universal TA cloning. Curr Issues Mol Biol. 2000;2(1):1–7.
5. Bhat S, Bialy D, Sealy JE, Sadeyen JR, Chang P, Iqbal M. A ligation and restriction enzyme independent cloning technique: an alternative to conventional methods for cloning hard-to-clone gene segments in the influenza reverse genetics system. Virol J. 2020;17(1):1–9.
6. Klock HE, Lesley SA. The Polymerase Incomplete Primer Extension (PIPE) method applied to high-throughput cloning and site-directed mutagenesis. In: High throughput protein expression and purification. Humana Press; 2009. p. 91–103.
7. Li MZ, Elledge SJ. Harnessing homologous recombination in vitro to generate recombinant DNA via SLIC. Nat Methods. 2007;4(3):251–6.

8. Bryksin AV, Matsumura I. Overlap extension PCR cloning: a simple and reliable way to create recombinant plasmids. BioTechniques. 2010;48(6):463–5.
9. Stevenson J, Krycer JR, Phan L, Brown AJ. A practical comparison of ligation-independent cloning techniques. PLoS One. 2013;8(12):e83888.
10. Aslanidis C, De Jong PJ. Ligation-independent cloning of PCR products (LIC-PCR). Nucleic Acids Res. 1990;18(20):6069–74.
11. Betton JM. High throughput cloning and expression strategies for protein production. Biochimie. 2004;86(9–10):601–5.
12. Park J, Throop AL, LaBaer J. Site-specific recombinational cloning using gateway and in-fusion cloning schemes. Curr Protoc Mol Biol. 2015;110(1):3–20.
13. Jewkes R, Sikweyiya Y, Morrell R, Dunkle K. Gender inequitable masculinity and sexual entitlement in rape perpetration South Africa: findings of a cross-sectional study. PLoS One. 2011;6(12):e29590.
14. Reece-Hoyes JS, Walhout AJ. Gateway recombinational cloning. Cold Spring Harb Protoc. 2018;2018(1):pdb-top094912.
15. Li MZ, Elledge SJ. MAGIC, an in vivo genetic method for the rapid construction of recombinant DNA molecules. Nat Genet. 2005;37(3):311–9.

Chapter 6
Fundamental Techniques of Recombinant DNA Transfer

Shriram Rajpathak, Rupali Vyawahare, Nayana Patil, and Aruna Sivaram

6.1 Introduction

In the year 1928, a classical experiment by Frederick Griffith using two different strains of *Streptococcus pneumoniae* played a very significant role in understanding the basics of bacterial transformation. The strains that he used were the non-virulent R strain and the pathogenic S strain which can be killed by exposure to higher temperature. In his experiment, Griffith observed that co-injecting the R and heat-killed S strain was lethal for mice. He could also isolate live R and S strain from the blood sample of the mice and concluded that the R strain was transformed into S strain by a "transforming principle" [1]. These sets of experiments proved that the trait for virulence was transferred from the S-type bacteria to the R-type cells, thus converting the non-virulent strain into a virulent one.

This work of Griffith was further studied by Avery et al. who discovered that the "transforming principle" was the nucleic acid, DNA [2].

The movement of genetic material between organisms is termed as horizontal gene transfer. In nature, this transfer happens mainly by three different mechanisms—conjugation, transformation, and transduction. Figure 6.1 gives a detailed description of these processes.

S. Rajpathak
ChAdOx1 and COVOVAX Vaccine Department, Serum Institute of India Pvt. Ltd., Pune, Maharashtra, India

R. Vyawahare
SGS Canada, Mississauga, ON, Canada

N. Patil · A. Sivaram (✉)
School of Bioengineering Sciences & Research, MIT ADT University, Pune, Maharashtra, India
e-mail: nayana.patil@mituniversity.edu.in; aruna.sivaram@mituniversity.edu.in

N. Patil, A. Sivaram, *A Complete Guide to Gene Cloning: From Basic to Advanced*, Techniques in Life Science and Biomedicine for the Non-Expert,
https://doi.org/10.1007/978-3-030-96851-9_6

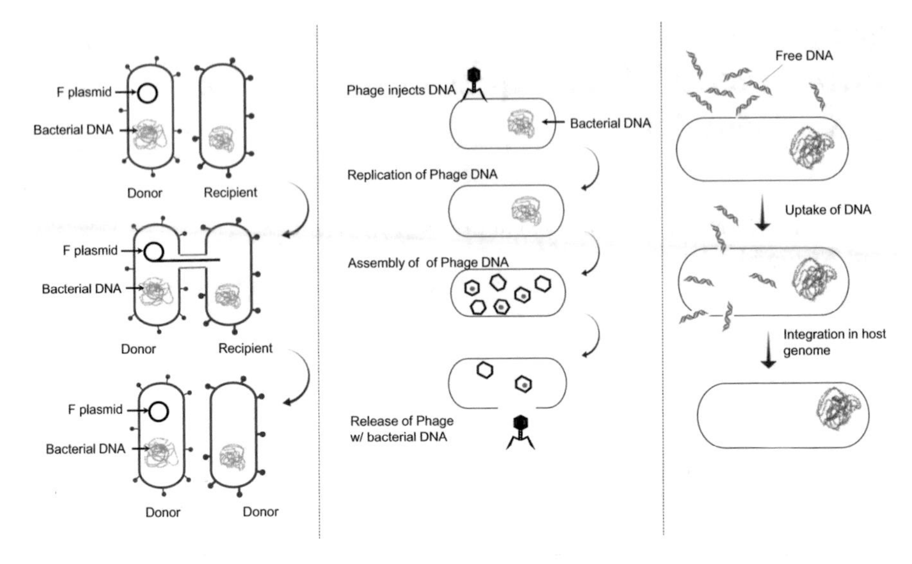

Fig. 6.1 Schematic representation of horizontal gene transfer by (**a**) Conjugation, (**b**) Transduction, and (**c**) Transformation. Conjugation involves transfer of genetic material when bacteria are in direct contact, while in transduction, the transfer happens through viruses. In transformation, the genetic material from the environment is taken up by bacterial cells

- Conjugation: Genetic material is transferred from the donor bacteria to the recipient bacteria when they are in direct contact. This genetic material is an episome which is a self-replicating unit and carries the F-factor which helps in establishing a contact between donor and recipient bacteria through formation of pilus.
- Transduction: Transfer of genetic material through viruses or viral vectors is called transduction. Genetic material from the bacteria gets incorporated into bacteriophages during its lytic or lysogenic life cycle and can be transferred from one bacterial cell to another.
- Transformation: As bacteria lyse, they shed genetic material like plasmid or fragmented DNA into the environment, which is taken up by adjacent living bacterial cells through a process called transformation.

Though initially discovered as a natural phenomenon occurring among prokaryotes, horizontal gene transfer has later been explored to deliver foreign DNA into the host organism as a part of RDT. Eventually, several host cells like fungi and mammalian cells were used in RDT. One of the most important steps in RDT is the delivery of the foreign DNA into the host cell. The transfer of foreign DNA deliberately into a eukaryotic cell, usually a mammalian cell, is called transfection. The success rate of rDNA transfer depends on various factors like the type, size, and purity of DNA, and medium and methods adopted for transfer [3]. To achieve successful transfer, the nucleic acid must get access to the cellular membrane, pass through the cytoplasmic compartment, and reach the interior of the nucleus. Before reaching the cell membrane, the nucleic acids must be stabilized, thus protecting it from degradation or deformation. The nucleic acid crosses the plasma membrane through penetration, adsorption, or ligand-mediated receptor binding and gets transported through the cytoplasm into the nucleus.

Transfection methods can be broadly classified into physical, chemical, and biological methods.

- Physical methods include electroporation, gene injection, laser beam, sonoporation, and use of magnetic nanoparticles.
- Chemical methods include transfection by lipids, peptides, cations, calcium chloride, etc.
- Biological methods involve the use of vectors to deliver the gene of interest into the host cells. This virus-mediated delivery of genes into a eukaryotic or prokaryotic cell is called transduction.

The method of DNA transfer will depend on various factors like the size and purity of the rDNA and the type of host cell. Selection of appropriate methods for DNA delivery is pertinent to ensure that it has been effectively transferred into the host cell. In this chapter, we will take a look at the important physical, chemical, and biological techniques for transfer of foreign DNA into prokaryotic and eukaryotic host systems.

6.2 Physical Methods of Gene Transfer

Several physical methods for transfer of DNA are widely used in prokaryotic and eukaryotic cells. These methods involve direct transfer of DNA into the cytoplasm or nucleus of the host cell. As no chemical or biological processes are involved, it is safer and consumes less time and labor. However, physical methods of gene transfer may be expensive as it requires specialized instruments which create a physical force that delivers the DNA into the host cell. Some of the common methods include microneedle injection, electroporation, biolistics, magnetofection, laser microbeams, ultrasound, and shock waves.

6.2.1 Microneedle Injections

This is one of the simplest and earliest methods of DNA transfer into the host cell and is considered to be minimally invasive. Here, the DNA is directly injected into the nucleus using a microneedle. Though considered to be a laborious process initially, development of automated injection systems has simplified the process and has made it reproducible and robust. Microneedles are fabricated from materials including stainless steel, ceramic, polymers, dextrin, etc. [4]. Various improvements and advancements in the system have widened the scope of its application.

Impalefaction, in which nanowires or nanofibers are used for DNA delivery instead of the needles is one such improvement. These advancements have worked toward improving the transfection efficiency while maintaining the viability of the cells.

6.2.2 *Biolistic Gene Transfer*

Established in 1987 for gene transfer in plant cells, biolistic method, also called gene gun method has been successfully applied in mammalian cell lines and animal models, bacteria, fungi, and algae. The nucleic acid is coated with high density carrier particles like gold, platinum, tungsten, etc. forming microspheres. Using high pressure inert gases or high voltage electric discharge, these microspheres are bombarded onto the target cell. This bombardment propels the nucleic acid into the target cell nuclei. The process requires fine optimization for efficient transformation [4]. Some of the parameters that affect the transformation efficiency include size and density of microspheres, the force of bombardment, the ratio of carrier particles to nucleic acid, the type and number of target cells and the temperature at which the process is being carried out. Figure 6.2 gives a schematic representation of the biolistic gene transfer method.

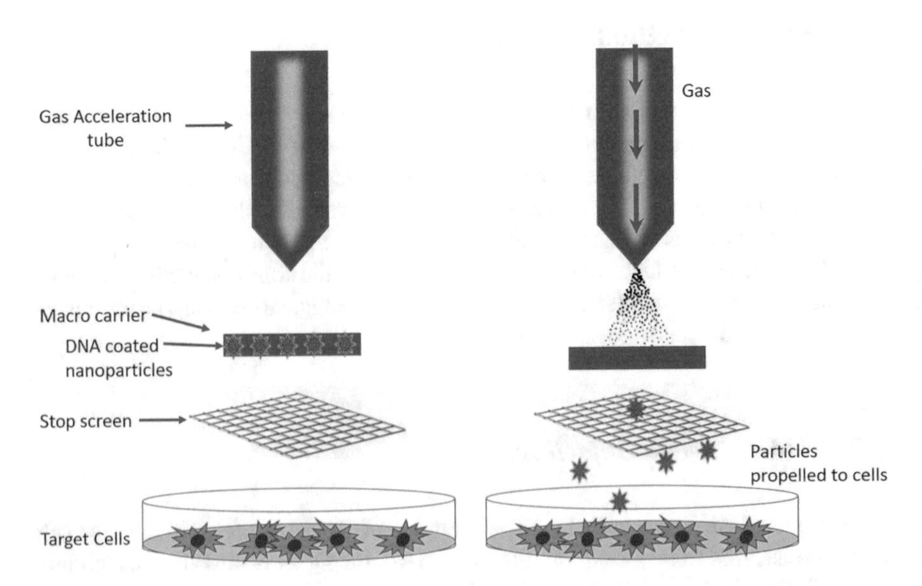

Fig. 6.2 Diagrammatic illustration of experimental design for gene transfer using Biolistic method. DNA coated on carriers are bombarded ballistically into target cells. The cells are allowed to recover thereby stabilizing the delivery of transgene

6.2.3 Electroporation

Since its discovery in 1982 by Neumann and colleagues, electroporation has evolved into one of the most versatile and successful systems for DNA delivery in prokaryotic and eukaryotic cells as well as in animal models [5]. Here, the target cell along with the nucleic acid is taken in a cuvette and placed in between two electrodes connected to electric supply. When a cell is exposed to a strong electric field, it results in formation of pores on the cell surface which allow nucleic acids to pass through. These pores are transient in nature and reseal without causing damage to the cell membrane structure or affecting the viability of the cell. Though initially used for transfecting plant protoplasts, electroporation has later found application in animal cells, fungi, bacteria, and even for animal models. Various factors associated with the electroporation condition, the host cells and the nucleic acid determine the efficiency of transformation. The electroporation conditions include pulse length, transmembrane potential, duration of electric field while the host-cell-associated parameters include the extent and duration of the permeated state. The global and local concentrations of the nucleic acid and form of DNA play an important role in transformation efficiency. Apart from these, tolerance of cells to membrane permeation and the heterogeneity of the cell population are also important factors. Nucleofection is a very significant improvisation made on electroporation and has been developed over the last decade for successful transfection of animal cell lines. This is a proprietary method developed by Amaxa, and owned by Lonza [6]. The electroporation conditions for each cell line are pre-programmed and are not revealed to the end user. This is one of the most successful methods for transfection of difficult-to-transfect primary cells, stem cells, and established cell lines. Gene transfer in animal models has been successfully achieved in skeletal muscles, liver, renal, dermal, pancreatic, intratumoral tissues, etc.

6.2.4 Sonoporation

Though ultrasound was in use for routine clinical diagnostic and therapeutic purposes since the 1960s, it gained momentum as a transfection technology in the 1990s when it was used to transfect chondrocytes and fibroblasts [7]. Sound waves create pores in cellular membranes and hence this method is called sonoporation. As with other physical methods, the underlying principle of this method also involves increasing the permeability of the cell membrane transiently. The nucleic acid to be transferred is taken in cell culture plates or flasks and ultrasound waves are applied through it. These ultrasound waves cause formation of bubbles in the media which contain the cells. These bubbles oscillate and subsequently collapse releasing energy which alters the permeability of the cell membrane, allowing passage of nucleic acid and other macromolecules through it. The transfection efficiency is influenced by various parameters like intensity of the ultrasound and its exposure time, the concentration of nucleic acid for transfer, the type of host cell and the temperature at which the procedure is carried out.

6.2.5 Laser-Microbeam-Based Transfection

Gene transfer using laser microbeams, also called optoporation, was established in the 1980s. Laser microbeams create pores on the cell surface, which allows the nucleic acid into the cells. Recently, several advancements have been made in the field which enable gene transfer into a particular site of interest or even a specific subcellular organelle by modifying the laser beam of appropriate wavelength and intensity. Figure 6.3 represents different physical methods like laser beam, sonoporation, and electroporation.

6.2.6 Magnetofection

Though initially studied for its potential application in targeted drug delivery, magnetofection has been used for gene delivery in cell lines and in animal models since early 2000s [8]. In this process, supramagnetic iron oxide particles are coated with the target nucleic acids. On application of a magnetic field, these nucleic acids come in very close proximity to the target cells and are taken into the cellular system, facilitated by endocytosis. For transfecting cell lines, the nucleic acid bound magnetic particles are mixed in cell culture media. This method is also used for delivery of nucleic acids to a particular target tissue in animal models as a part of gene therapy studies [9]. The nucleic acid in complex with the magnetic particle is

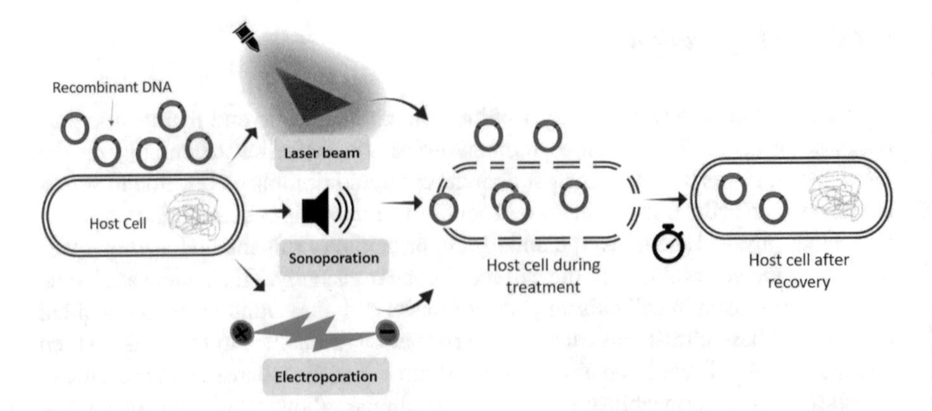

Fig. 6.3 Schematic representation of gene delivery mediated by laser beam, sonoporation, and electroporation. These methods transiently form pores in the holes in the cell membrane which facilitates the uptake of DNA fragments

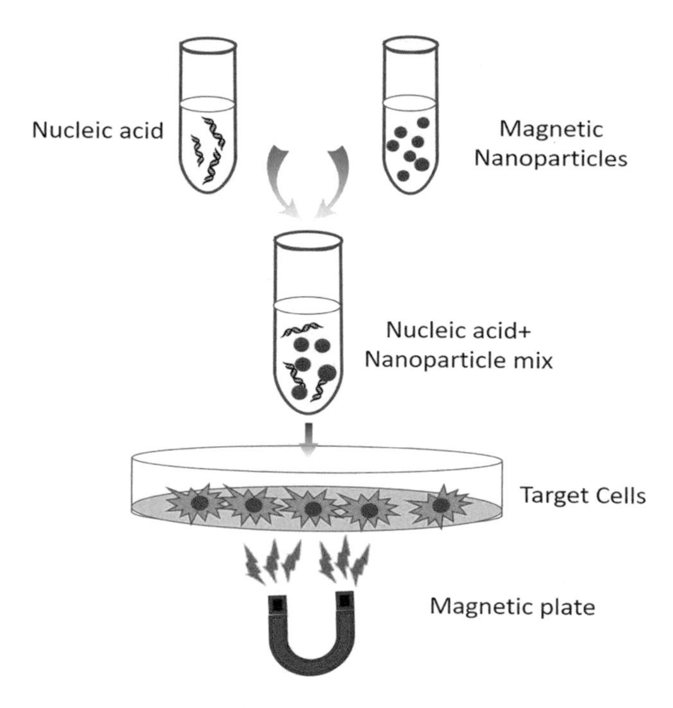

Fig. 6.4 Schematic diagram of gene delivery mediated by magnetic field. Nucleic acids are linked to magnetic nanoparticles and are attracted toward the surface of the target cell under an external magnetic field. The complex is internalized by the process of endocytosis into the cells

administered through an intravenous route and is captured at the specific site or tissue using strong, high gradient magnets externally. The transfection efficiency depends on the type of cells, the flux density and gradient of the magnet, the contact time between the nucleic acid and the cells, the flow rate of blood, etc. Figure 6.4 gives a schematic diagram of gene delivery mediated by magnetic field.

A summary of the physical methods for gene transfer is described in Table 6.1.

6.3 Chemical Methods of Gene Transfer

As discussed in the previous section, physical methods are easy to use for almost all types of cells and animal models. However, the physical methods may require sophisticated instruments and skilled manpower. It may also result in damaging the host cell or the target nucleic acid. The chemical methods overcome the drawbacks of physical methods to a certain extent. Cationic polymers, cationic lipids, calcium phosphate, etc. are used in chemical transfection. The positively charged chemicals form a complex with negatively charged nucleic acids. These complexes are attracted toward the negatively charged cell membrane and pass through it [10].

Table 6.1 Physical methods used for gene transfer

Method	Brief description	Advantages	Disadvantages
Microneedle injection	Directly injected into the nucleus using a microneedle	High transfer efficiency	Laborious process as only a single cell can be injected at a time
		Direct delivery of DNA into the nucleus minimizes the contact with cytoplasmic nucleases, thus reducing degradation	Requires sophisticated tools and computer aided techniques for precise delivery of DNA
Biolistic gene transfer	Using high pressure inert gases or high voltage electric discharge, DNA microspheres are bombarded onto the target cell	Delivery of nucleic acid into a wide variety of target cells is possible	High cost involved in the instrumentation and preparation of microspheres
		Multiple transgenes can be used	Requires fine optimization for best transfection efficiency
Electroporation	Cells are exposed to strong electric field, which forms pores on the cell surface and allow nucleic acids to pass through	Less damage to cell structure and viability	Protocol may require extensive standardization
		Simple, involves lesser time and cost	Area that can accommodate the cells between the electrodes is limited; this limits the number of cells that can be transfected
Sonoporation	Cells are exposed to ultrasound waves which forms pores on the cell surface and allow nucleic acids to pass through	Involves moderate cost	Only modest efficiency of transfection has been observed
		Simple noninvasive method for gene transfer	If the parameters are not optimized carefully, it may result in damage of cell membrane
Laser-microbeam-based transfection	Laser microbeams create pores on the cell surface, which allows the nucleic acid into the cells	Noninvasive, simple method of transfection	The method involves high cost in the form of instrumentations
		Targeted gene delivery to specific sites or tissues can be done	The method may result in low efficiency due to limited penetration of laser beam
Magnetofection	Magnetic field enables iron oxide-coated nucleic acids to gain cellular entry through endocytosis	Guided delivery of nucleic acids to specific sites in animal models	Requires formulation of magnetic nanoparticles with the nucleic acid
		One of the noninvasive techniques	Strong external magnets and fine tuning of the transfection parameters is required for good transfection efficiency

Some important chemical transfection methods are discussed below. Figure 6.5 gives an overview of chemical approaches for gene delivery to cells.

Box 6.1The most commonly used method in laboratories worldwide is to prepare competent bacterial cells using calcium chloride and transfer the foreign nucleic acid through heat shock. Bacterial cells which can take up the foreign nucleic acid molecule are called competent cells. Though some bacteria may be competent naturally, for gene transfer in laboratory conditions, bacterial cells are converted into competent cells through treatment with calcium chloride. ***E. coli*** **is the most common bacterial species used in rDNA technology. The bacterial cells are harvested from liquid culture in the mid-log phase and are incubated in CaCl2 in order to make the cell membrane permeable. Competency can be enhanced by replacing or substituting calcium ions with other cations like potassium or rubidium ions. The foreign nucleic acids are then transferred to these cells through a heat shock method. In this method, the cells and DNA are mixed and incubated on ice for 5–45 mins. A heat shock for 25–45 s at 37–42 °C is applied onto the cells followed by incubation in ice again for 2–5 min. The heat shocked cells are allowed to recover in antibiotic free media and later cultured in appropriate media and screened for appropriate colonies**

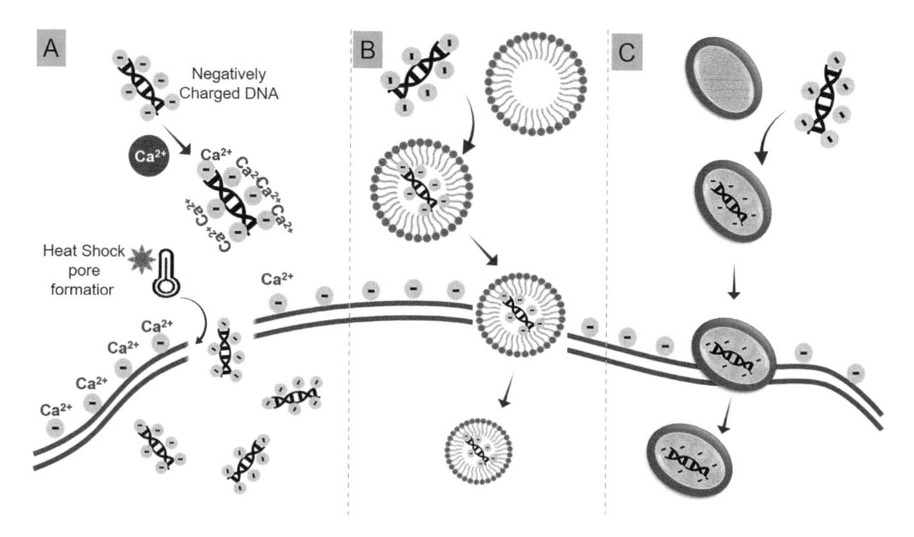

Fig. 6.5 Overview of chemical approaches for gene delivery to cells. (A) $CaCl_2$ methods (B) Polymer-based methods involving carriers such as chitosan, PEI, and dendrimers. (C) Liposomes method

6.3.1 Calcium Phosphate Precipitation

This is one of the widely used, cost effective techniques for gene transfer. The nucleic acid-calcium phosphate co-precipitate formed by mixing nucleic acid with calcium chloride in the phosphate buffer is incubated with the target cells. This co-precipitate is phagocytosed into the cells and reaches the nucleus. Formation of the co-precipitate is one of the critical steps in this method and it is dependent on concentration of Ca^{2+} ions, phosphate ions, nucleic acid and the temperature and pH of the reaction system. These factors also influence the transfection efficiency in a profound way. Several researchers have developed nanocomposites of calcium phosphate which encapsulates nucleic acids. These nanocomposites exhibited higher transfer efficiency than the conventional co-precipitates.

6.3.2 Liposomes

Liposome-based gene transfer system is one of the most commonly used among the chemical methods. Liposomes are bilayered synthetic membrane structures composed of lipids. When nucleic acid and lipids are mixed in a certain ratio, the lipid molecules form a bilayer around the nucleic acids. These liposomes which encapsulate the nucleic acid interact with the host cell and enter it [11, 12]. The formation of liposome-nucleic acid complex is one of the important steps; it depends on various factors like the conditions at which the procedure is carried out, the ratio between the nucleic acid and the lipids, ionic strength of the buffer, etc. Liposomes provide a stable environment to the nucleic acids and protect it from enzymatic degradation. Several cationic lipid molecules have been studied for their potential in gene transfer. One of the significant improvements of this method is coating of the liposomal surface with poly ethylene glycol (PEG). This coating facilitates gene transfer in animal models by stabilizing the liposomes and improving its circulation time in the blood. This also minimizes the aggregation of liposomes and improves the transfer efficiency by facilitating better distribution across tissues.

6.3.3 Polymeric Carriers

Cationic carriers are conjugated to the target gene for transfer and are taken up through endocytosis by the host cells. Use of these carriers improves the transfer efficiency as compared to liposomes. Several polymeric carriers are used in this technique [13].

(a) Chitosan: It is one of the most commonly used polymers which shows biocompatibility, safety, and low toxicity. It forms a complex with nucleic acid, the stability of which is determined by length of the chitosan chain and the ratio between chitosan and the nucleic acid. The efficiency of gene transfer using this method depends on the molecular weight of chitosan, the cell type that is used and the physical and chemical properties of the chitosan–nucleic acid complex. Chitosan is insoluble in physiologic pH, thus limiting its use for potential applications. Optimization of parameters and modification of hydrophilic and hydrophobic properties are important for its effective utilization [14].

(b) Polyethylenimine (PEI): It is an inexpensive and simple reagent used for gene transfer. It has several unique properties like high capacity for nucleic acid condensation and intrinsic endosomal activity. The polymer forms nano-complexes with the nucleic acid [15]. The efficiency of gene transfer depends on the ratio between the polymer and the nucleic acid, molecular weight of PEI. Better transfection efficiency has been observed in the presence of higher amounts of free PEI. PEI with higher molecular weight has higher cationic charge and hence condenses the nucleic acid better, forming smaller nanoparticles. However, it has been observed that high molecular weight PEI is cytotoxic to the cells and may lead to cell apoptosis. In animal models, PEI has been shown to be aggregating red blood cells, and is not easily metabolized and excreted. Introduction of hydrophobic chains and substituents have improved the transfection efficiency.

(c) Polyamidoamine Dendrimers: These are uniformly dispersible macromolecules with highly branched spherical polymers with a surface made up of primary amino group. They are highly hydrophilic in nature. Dendrimers form a stable and highly soluble complex with nucleic acid [16]. Synthesis of dendrimers is laborious and hence these polymers are expensive. The synthesis occurs by addition of successive branching groups to a core group. Each layer is called a generation. Transfection efficiency depends on the generation of dendrimers. At lower generations the transfection efficiency is also low. At higher generations, though transfection is better, the cytotoxicity also increases. To overcome this drawback, several modifications are being made by researchers to improve the transfection efficiency in lower generations.

(d) Poly (lactide-co-glycolide): This polymer constitutes of lactic acid and glycolic acid linked together through ester bonds. PLGA protects the encapsulated nucleic acid from enzymatic degradation in animal cells. However, the rate of release of the encapsulated nucleic acid is low. PLGA also possesses negative charges which interfere with the encapsulation of nucleic acid. The polymer also has an acidic microenvironment which affects the properties of the nucleic acid. The low release rate also contributes toward lower gene transfer efficiency.

A summary of the chemical methods for gene transfer is described in Table 6.2.

Table 6.2 Chemical methods used for gene transfer

Method	Brief description	Advantages	Disadvantages
Calcium phosphate precipitation	Co-precipitate formed by mixing nucleic acid with calcium chloride in phosphate buffer is phagocytosed by the cells	Involves biocompatible and biodegradable materials	Modest transfection efficiency and low reproducibility
		Easy to handle, cost effective method	
Liposomes	Liposome encapsulates the nucleic acid, interacts with the host cell and enters it	Biocompatible and less immunogenic	Lesser cell adhesion and lower transfection efficiency
		Can be used for gene delivery in vitro and in vivo	Have potential cytotoxicity, thus limiting their use in gene therapy
Chitosan	Forms a complex with nucleic acid and enters the cell	Biocompatible and biodegradable	Low solubility in physiological pH
		Polymer offers low immunogenicity and is less toxic	Low efficiency of gene transfer
Polyethylenimine	Polymer forms nano-complexes with the nucleic acid	High capacity to condense the nucleic acid resulting in smaller nanoparticles	Poor metabolism and excretion from the body
		Intrinsic endosomal activity	May cause cytotoxicity
Polyamidoamine dendrimers	Form a stable and highly soluble complex with nucleic acid	Higher solubility, which helps in release of the nucleic acid molecule	High cost
		Provides mono-dispersion	Fine optimization required to achieve satisfactory transfection with least cytotoxicity
Poly (lactide-co-glycolide)	Protects the encapsulated nucleic acid from enzymatic degradation in animal cells	Safe and biodegradable	Low transfection efficiency
			Low release rate
			Acidic nature of the polymer may affect the properties of nucleic acid

6.4 Biological Methods of Gene Transfer

Biological methods are one of the most advanced gene delivery methods. They have high efficiency and specificity. The major techniques involved in this method are described below. Figure 6.6 represents the important biological methods.

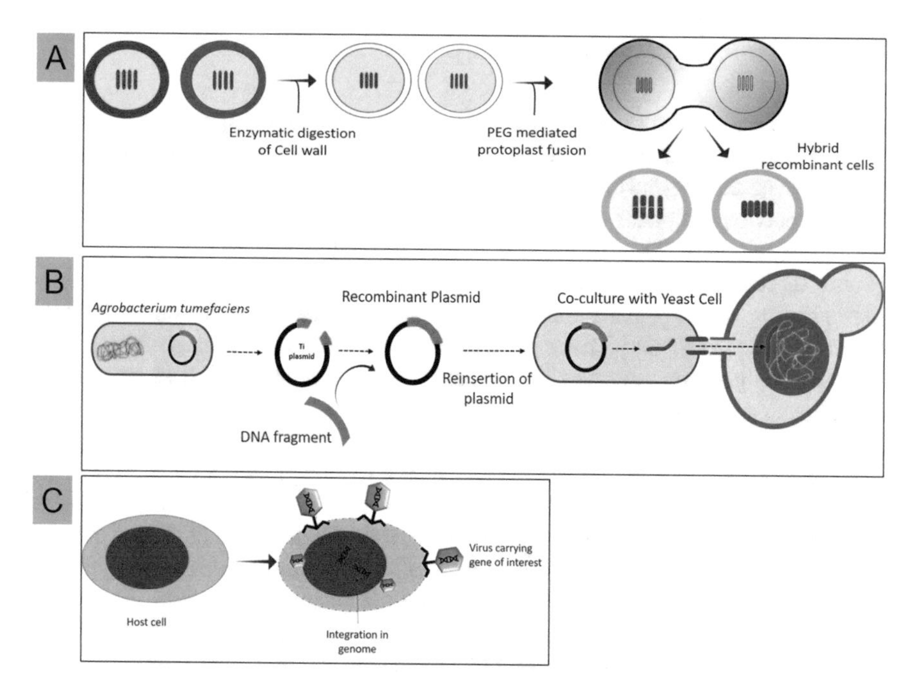

Fig. 6.6 Schematic diagram showing major steps involved in transfer of DNA using biology methods. (**a**) Protoplast fusion: Enzymatic digestion of fungal cells wall exposes the protoplast and undergo fusion mediated by PEG resulting in recombined hybrids. (**b**) *Agrobacterium*-mediated gene transfer: The Ti plasmid is modifies so as to carry the gene of interest. *Agrobacterium* carrying the recombinant plasmid transfers the DNA into fungal cells generating transgenic cells. (**c**) Virus mediated: The gene of interest is directly inserted into a virus which is genetically engineered to deliver the gene into the host cell

6.4.1 Bacteria-Mediated Gene Transfer

This method also called bactofection uses bacteria to direct the gene into the host cell. Several bacterial strains like *Listeria monocytogenes, Salmonella typhimurium, Salmonella choleraesuis* are used. In this process, bacterial strains are transformed with plasmids of interest. The bacterial strains carrying the plasmid are then transferred to the target cell. The bacterial cell is lysed in the cytoplasm of the target cell, releasing the plasmid which enters the nucleus of the target cells. Several methods like using integrin receptors and lipofectamine-mediated bactofection enhances the transfer efficiency. This method has potential applications in DNA vaccination and in gene therapy [17].

6.4.2 *Protoplast Fusion*

Protoplasts are formed by removing the cell wall of a cell through enzymatic treatment. Plants are treated with cellulase or pectinase, bacteria with lysozyme and fungi with glucanase and chitinase for the removal of cell walls. In protoplast fusion technique, two protoplasts which are genetically different are fused together to form a single protoplast. This parasexual hybrid protoplast consists of fused nuclei. Protoplast fusion may occur as a spontaneous phenomenon between adjacent protoplasts during the enzymatic treatment. However, spontaneous fusion is not very common as isolated protoplasts carry a negative charge and repel each other. Hence, for effective gene transfer, the fusion between the protoplasts has to be induced using mechanical, chemical, or electrofusion methods. In mechanical methods, a micromanipulator or perfusion micropipette is used to bring the isolated protoplasts into close proximity. Chemofusogens like poly ethylene glycol or calcium ions are used to bring the protoplasts together which results in their agglutination and fusion in chemical fusion technique. In electrofusion, mild electric current passed through capillary glass electrodes placed in contact with the protoplasts causes the fusion. The transfection efficiency largely depends on the extent of protoplast fusion. The factors that affect protoplast fusion include viability of protoplasts after enzymatic treatment of the cells, in vitro culture like media and supplements, genotype of the organism, environment and culture condition of the protoplast source, etc. [18]. This is a very important technique in somatic hybridization and is most widely used for gene transfer in plants and fungi.

6.4.3 Agrobacterium-*Mediated Gene Transfer*

Agrobacterium tumefaciens is a plant pathogen which causes the crown gall disease. The pathogen harbors the tumor inducing (Ti) plasmid which has two very essential genetic components—the T-DNA and the Vir genes, both of which are present in the tumor inducing (Ti) plasmid. As the pathogen enters through the wounded region, the plants release phenolic acetosyringone which enhances the binding between the pathogen and the host. The acetosyringone activates VirA protein of the bacteria which triggers a series of signaling events, leading to the formation of a single stranded copy of T-DNA. To this strand, several Vir proteins bind, forming the T-complex. The nuclear target signal produced by the Vir proteins guides the T-complex into the nucleus, where it gets integrated into the genome of the host cell. This method has been successfully adopted in the laboratory to deliver genes into plant cells. The oncogene present in the pathogen is replaced by the gene of interest for transfer of genes. The integration of the DNA is followed by regeneration of a complete plant from the transfected plant cells. The transfection efficiency is affected by various factors like the species and cell density of plants, the growth regulators and media, the environmental factors and bacterial strain. Appropriate optimization of these parameters is important to yield good gene transfer efficiency. This method has been used to synthesize several pharmaceutically relevant products in plants, to improve the resistance of plants toward pests, to improve tolerance against biotic and abiotic factors, etc. [19].

6.4.4 Viral Vector-Based Gene Transfer

Here viruses are used as gene carriers. The method is also called transduction. A high rate of replication and protein expression make viruses a suitable carrier to deliver the genes in a host cell. The virus particle in which the gene of interest is packaged enters the host cell through a receptor-mediated process. This delivery system has found application in gene therapy. Common viral vectors used for gene delivery include adenoviruses and lentiviruses. A detailed explanation for adenoviral vectors and lentiviral vectors has been provided in Chap. 2. Other viral vectors include retroviral vectors, Herpes simplex viral vectors, pox viral vectors. The viruses used as delivery agents should be genetically engineered so that they can deliver the genes without causing any pathogenic effects on the host cells.

A summary of the biological methods for gene transfer is described in Table 6.3.

Table 6.3 Biological methods used for gene transfer

Method	Brief description	Advantages	Disadvantages
Bacteria-mediated gene transfer	Uses bacteria to direct the gene into the host cell	Simple method	May cause adverse events
		High transfer efficiency	Genetically modified bacteria which lyses within the host cell may have to be used
Protoplast fusion	Two protoplasts which are genetically different are fused together to form a single protoplast	Binary vector is not required for this method	Tedious and laborious process
		Can be used for most plant species, shows high efficiency of gene transfer	Multiple fusion and regenerative steps may be required
Agrobacterium-mediated gene transfer	Transfer of genes through *agrobacterium*; the oncogene is replaced by gene of interest and integrates with the host genome	Inexpensive and easy to use	Narrow host range, limited number of plants can be transfected using this method; cannot transfer large DNA
		Low copy number DNA can be inserted into the host cells	Plant regeneration protocols have to be improved

6.5 Summary

Various biological, chemical, and physical methods are available for delivering the gene into target cells. Depending on the target cells the utilization of method and thus efficiency of method differs. Various efforts are being carried out to understand the behavior of cells under different conditions, how plasma membranes of organisms differ in composition, how nucleic acids are transported to the cell environment and to the nucleus. An effective answer to these questions will help to invent future advanced technologies for transfection of cells.

Self assessment

Q1. A scientist has to transfer genes to a plant cell. Write any two methods that can be used for this

A1. The gene transfer in plants can be performed using protoplast method or *Agrobacterium*-mediated transfer. In the protoplast fusion method, two protoplasts are brought to proximity and are fused using physical, chemical, or electrofusion methods. This results in the formation of genetic hybrids of plants. In *Agrobacterium*-based gene transfer, the oncogene of the *Agrobacterium* is replaced with the gene of interest. When this bacterium enters into the host cell, it releases the genetic material into the nuclei, where it gets integrated along with the host genetic material. This host cell can be regenerated into a complete plant.

Q2. Mention a physical method which uses electric current for transfer of genes

A2. Electric current is used for gene transfer in a method called electroporation. The target cell along with the nucleic acid is taken in a cuvette and placed in between two electrodes connected to electric supply. When a cell is exposed to a strong electric field, it results in formation of pores on the cell surface which allow nucleic acids to pass through. These pores are transient in nature and reseal without causing damage to the cell membrane structure or affecting the viability of the cell.

Q3. How are liposomes used for gene transfer?

A3. Liposome-based gene transfer system is one of the most commonly used among the chemical methods. Liposomes are bilayered synthetic membrane structures composed of lipids. When nucleic acid and lipids are mixed in a certain ratio, the lipid molecules form a bilayer around the nucleic acids. These liposomes which encapsulate the nucleic acid interact with the host cell and enter it.

References

1. Griffith F. The significance of pneumococcal types. Epidemiol Infect. 1928;27(2):113–59.
2. Avery OT, MacLeod CM, McCarty M. Studies on the chemical nature of the substance inducing transformation of pneumococcal types induction of transformation by a desoxyribonucleic acid fraction isolated from pneumococcus type III. J Exp Med. 1944;79(2):137–58.

3. Liu X, Liu L, Wang Y, Wang X, Ma Y, Li Y. The study on the factors affecting transformation efficiency of E. coli competent cells. Cell. 2014;27(3 Suppl):679–84.
4. Alsaggar M, Liu D. Physical methods for gene transfer. Adv Genet. 2015;89:1–24.
5. Neumann E, Schaefer-Ridder M, Wang Y, Hofschneider PH. Gene transfer into mouse lyoma cells by electroporation in high electric fields. EMBO J. 1982;1:841–5.
6. Freeley M, Long A. Advances in siRNA delivery to T-cells: potential clinical applications for inflammatory disease, cancer and infection. Biochem J. 2013;455(2):133–47.
7. Kim HJ, Greenleaf JF, Kinnick RR, Bronk JT, Bolander ME. Ultrasound-mediated transfection of mammalian cells. Hum Gene Ther. 1996;7:1339–46.
8. Widder KJ, Senyei AE. Magnetic microspheres: a vehicle for selective targeting of drugs. Pharmacol Ther. 1983;20:377–95.
9. Sizikov AA, Kharlamova MV, Nikitin MP, Nikitin PI, Kolychev EL. Nonviral locally injected magnetic vectors for in vivo gene delivery: a review of studies on Magnetofection. Nano. 2021;11(5):1078.
10. Kim TK, Eberwine JH. Mammalian cell transfection: the present and the future. Anal Bioanal Chem. 2010;397(8):3173–8.
11. Tenchov R, Bird R, Curtze AE, Zhou Q. Lipid nanoparticles-from liposomes to mRNA vaccine delivery, a landscape of research diversity and advancement. ACS Nano. 2021;15(11):16982–7015.
12. Large DE, Abdelmessih RG, Fink EA, Auguste DT. Liposome composition in drug delivery design, synthesis, characterization, and clinical application. Adv Drug Deliv Rev. 2021;176:113851.
13. Jin L, Zeng X, Liu M, Deng Y, He N. Current progress in gene delivery technology based on chemical methods and nano-carriers. Theranostics. 2014;4(3):240–55.
14. Cao Y, Tan YF, Wong YS, Liew MWJ, Venkatraman S. Recent advances in chitosan-based carriers for gene delivery. Mar Drugs. 2019;17(6):381.
15. Longo PA, Kavran JM, Kim MS, Leahy DJ. Transient mammalian cell transfection with polyethylenimine (PEI). Methods Enzymol. 2013;529:227–40.
16. Palmerston Mendes L, Pan J, Torchilin VP. Dendrimers as Nanocarriers for nucleic acid and drug delivery in cancer therapy. Molecules. 2017;22(9):1401.
17. Oladejo M, Paterson Y, Wood LM. Clinical experience and recent advances in the development of listeria-based tumor immunotherapies. Front Immunol. 2021;12:642316.
18. Ahmed AA, Miao M, Pratsinakis ED, Zhang H, Wang W, Yuan Y, Lyu M, Iftikhar J, Yousef AF, Madesis P. Protoplast isolation, fusion, culture and transformation in the Woody Plant Jasminum spp. Agriculture. 2021;11(8):699.
19. Peng LH, Gu TW, Xu Y, Dad HA, Liu JX, Lian JZ, Huang LQ. Gene delivery strategies for therapeutic proteins production in plants: emerging opportunities and challenges. Biotechnol Adv. 2021;54:107845.

Chapter 7
Selection, Screening, and Analysis of Recombinant Clones

Madhura Chandrashekar, M. S. Maralappanavar, Premjyoti Patil, Nayana Patil, and Aruna Sivaram

7.1 Introduction

In the previous chapters, we have discussed a wide range of vectors and the different techniques applied for cloning and transformation in the case of bacteria, fungi, plant, and animal cells. In this circumstance screening and analysis of recombinants are the critical steps that influence the success of the cloning experiment. For example, the genomic and cDNA library contains thousands of clones, wherein each clone harbors different DNA fragments or genes. It is essential to identify the clones carrying the desired gene in a single shot to save time, effort, and make the process economical.

In this line, the current chapter details the different selection and screening methods that can be employed to identify transformants, recombinant clones and various strategies to select the clones with the gene of interest among the recombinant clones. By the end of the chapter, further analysis of the identified clones has been discussed.

Before going into the actual method, it is essential to know about the two terminologies used frequently in this chapter i.e., selection and screening. Here selection refers to picking out only the transformed cells or cells containing vectors (which

M. Chandrashekar · N. Patil · A. Sivaram (✉)
School of Bioengineering Sciences & Research, MIT ADT University, Pune, Maharashtra, India
e-mail: madhura.chandrashekar@mituniversity.edu.in; nayana.patil@mituniversity.edu.in; aruna.sivaram@mituniversity.edu.in

M. S. Maralappanavar
University of Agricultural Sciences, Dharwad, Karnataka, India

P. Patil
Basaveshwar Engineering College, Bagalkote, Karnataka, India

N. Patil, A. Sivaram, *A Complete Guide to Gene Cloning: From Basic to Advanced*, Techniques in Life Science and Biomedicine for the Non-Expert,
https://doi.org/10.1007/978-3-030-96851-9_7

may contain genes or may not) from the non-transformants (cells without any vector) after transforming the host cells with the desirable vector. It is achieved by exerting pressure under certain defined conditions (like the addition of antibiotics in the selecting medium), where only the clones carrying the vector can grow on the media as vectors express the desirable trait (like antibiotic resistance) to defy the condition. Further screening among the transformants is done to separate the recombinant clones (transformants harboring vectors with foreign DNA segment) from the nonrecombinant clones (transformants carrying only vectors without any foreign DNA segment in them). In the case of the library (genomic/cDNA) the recombinant clones are screened additionally to identify the clones carrying the gene or DNA segment of interest. The screening methods applied need to be highly sensitive and should have scope for automation so that a large number of recombinant clones can be screened in less time and in a single step. Such methods are beneficial to screen the genomic or cDNA library for the desired clones [1, 2].

7.2 Selection of Transformants

The selection of transformants is the first and foremost step in a cloning experiment wherein transformed cells (with vector) are differentiated from the non-transformed cells (without vector). It is achieved by selecting for the trait expressed by the vector under peer selection pressure. For example, the case of the most commonly used pUC plasmid cloning vector consists of the ampicillin resistance gene. When the cells transformed with this plasmid vector are plated on the Luria agar media containing ampicillin, only the cells that received the plasmid vector survive due to the expression of ampicillin antibiotic resistance genes present in the plasmid (Fig. 7.1). The cells without the plasmid vector do not survive as they do not harbor the ampicillin antibiotic resistance gene. Similarly, in the case of fungi, the transformed and non-transformed cells can be differentiated with the help of a few auxotrophic marker genes, which encode critical enzymes for amino acid synthesis.

The auxotrophic markers can only be used if a corresponding auxotrophic mutant host is available for example, the use of the Trp1 auxotrophic gene in a vector, which encodes for the tryptophan amino acid. The corresponding host should be an auxotrophic mutant for the amino acid tryptophan. This means the host cell is incapable of growing on the minimal media and its growth can be achieved by enriching the media with the tryptophan. If such host cells are transformed with the vector containing the Trp1 gene, the transformed cells can grow on the minimal media, and cells without the vector are incapable of growing (Fig. 7.2).

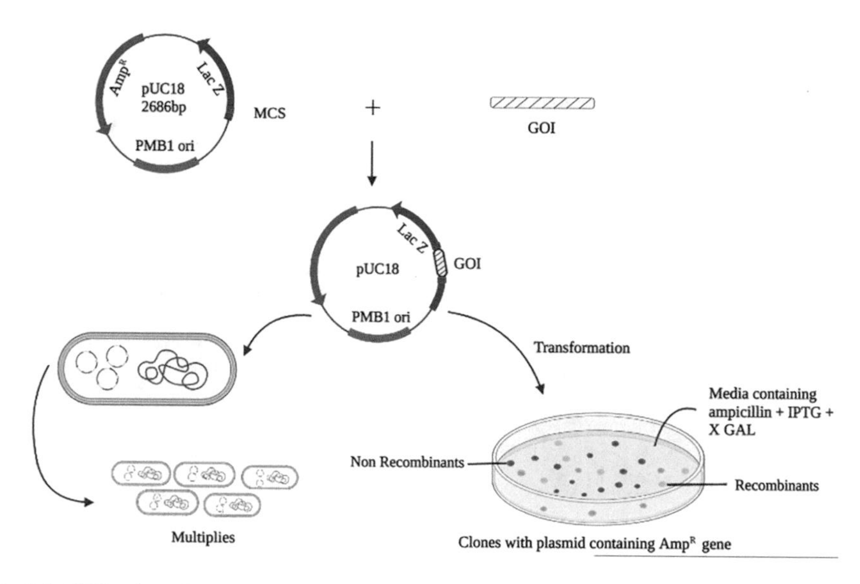

Fig. 7.1 Utilization of Antibiotic selection marker present in the plasmid vector (pUC18) to distinguish the transformants from non-transformants (*GOI* Gene of interest)

7.3 Screening of Recombinant Clones

During cloning, first, the vector and the insert are cut with the same restriction enzyme. The two restriction digestion products are ligated using the enzyme DNA ligase. There are chances of obtaining three different products during the process like self-ligated vector (vector without gene), vector with the gene of interest (recombinants), and gene–gene ligated product. When these ligated products are transformed into the host systems and plating is done on the agar medium. The plate consists of a mixture of clones for all the above-ligated products (Fig. 7.3). It is necessary to deploy the screening technique to save time, resources, and effort, where recombinants can be easily differentiated from nonrecombinants.

Instead of using a single enzyme to cut the insert and the vector, two different enzymes can be used. Thus the two ends of the vectors are cut with two different enzymes, and the same two sets of enzymes are used to cut the two ends of the insert. After ligation, there are chances of obtaining mainly the recombinant vectors. This method is commonly referred to as directional cloning.

Generally, recombinant clones are identified either by hybridizing the insert sequence using labeled probes or by inactivating one of the genes encoded by the vector. Gene inactivation is the most commonly used straightforward simple technique to identify recombinant clones. It is achieved by knocking out the gene encoded by the vector by inserting the foreign DNA within its coding sequence. This method can be further classified into two subtypes i.e., negative and positive selection.

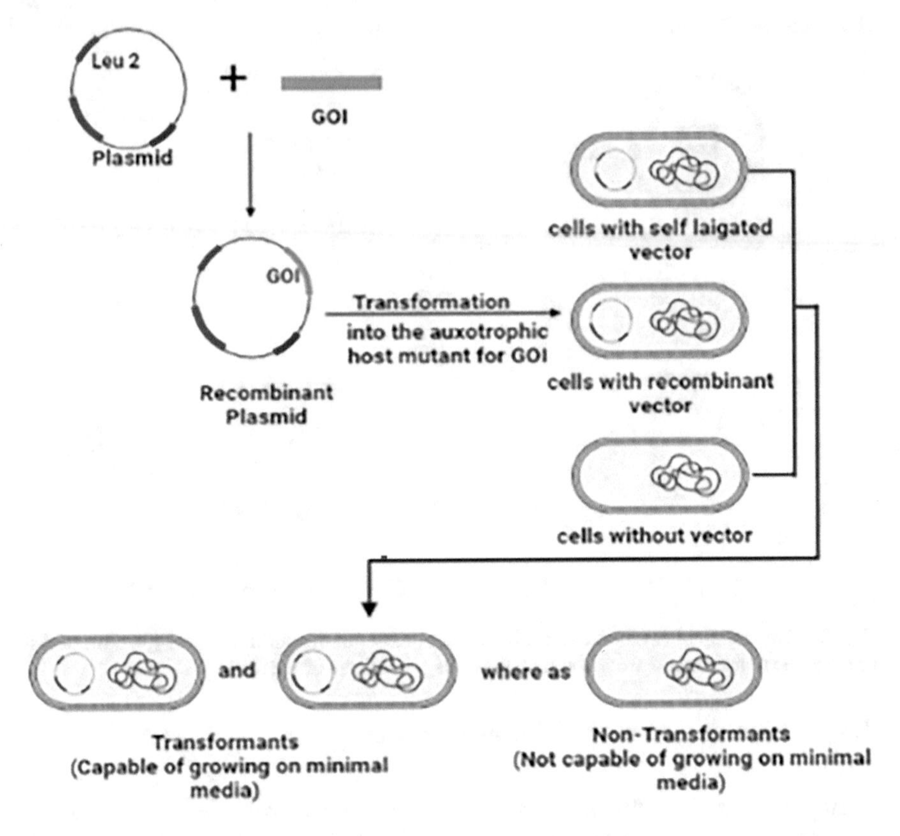

Fig. 7.2 Utilization of the auxotrophic markers which are part of the vector system to select only the cells receiving the vectors (transformants) in the case of yeast cells

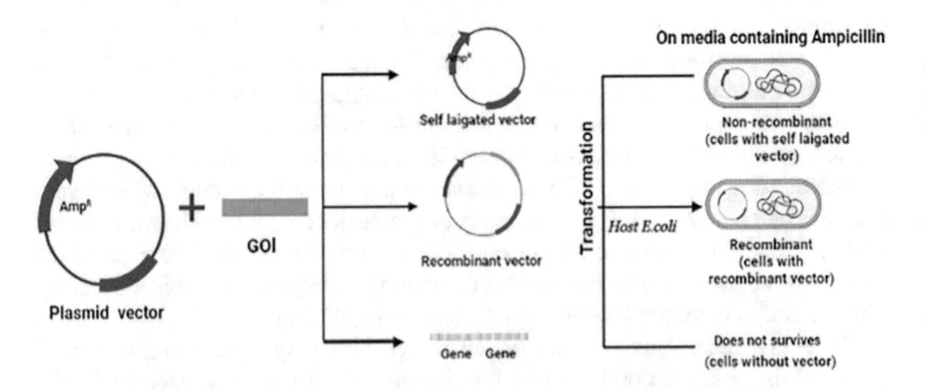

Fig. 7.3 Possible products produced after the transformation of the vector into a host system and the chance of their survivability on the media containing antibiotics

7.3.1 *Negative Selection*

In this technique, the selection is made for the loss of gene product encoded by the vector. Here, the vector genes are inactivated by inserting foreign DNA segments into the vector. This can be accomplished either by the indirect (replica plating) or direct (chromogenic) approach. Let us discuss each method in detail.

7.3.1.1 Antibiotic Selection by Replica Plating

In this technique, the recombinants are selected by screening the transformants for the loss of gene function encoded by the vector.

For example, the pBR322 plasmid vector encodes for two antibiotic resistance genes i.e., ampicillin and tetracycline. Thus the cells transformed with pBR322 vectors can grow on media containing both the antibiotics as mentioned above. If the gene of interest is inserted within any one of these antibiotic resistance genes, the recombinant clones remains tolerant to one antibiotic and sensitive to the other antibiotic in which the foreign DNA segment was inserted. As illustrated in Fig. 7.4 the tetracycline gene can be inactivated by inserting the foreign DNA sequence within. The recombinant clones are capable of multiplying on media containing ampicillin but not on tetracycline. The recombinant clones do not survive on the media containing tetracycline, they cannot be selected directly by culturing them on the tetracycline plate. Hence the recombinant clones can be selected by indirect means. The

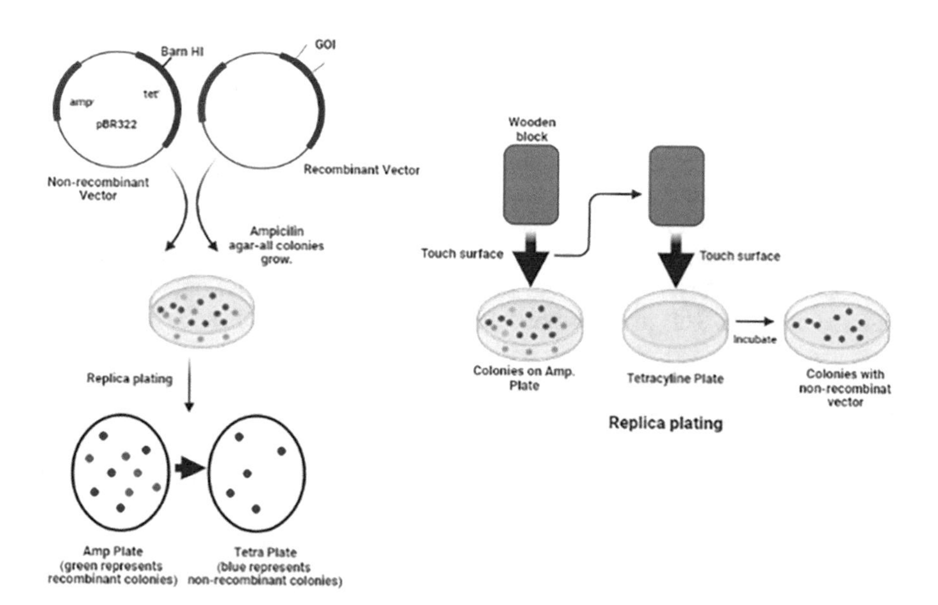

Fig. 7.4 Antibiotic selection by replica plating to detect the recombinant clones

replica plating technique can attain this. The transformed cells are plated on a media containing ampicillin (master plate) first, and then an impression of the colonies is traced by keeping the nitrocellulose membrane on the plate. These membranes are placed on the plate containing both ampicillin and tetracycline (replica plate) there by the position of the colonies remains the same in both the plates. In presence of both ampicillin and tetracycline, only the nonrecombinant clones can multiply on the plate due to the inactivation of tetracycline gene. The recombinant clones can be selected by tracing back to the master plate for the same position where there is no growth in replica plate.

Advantages

1. Here the selection is based on the presence or absence of colonies, which makes the method simpler.

Limitations

1. When both ampicillin and tetracycline are present, transformant growth is less than media containing ampicillin or tetracycline alone. This additive effect implies that ampicillin plus tetracycline confers a more substantial selective pressure on the cells, lowering the survival rate.
2. Replica plating of the master plate is required to retrieve the target cones, which is cumbersome.

7.3.1.2 Chromogenic Method

It is one of the most widely and commonly used techniques, where the recombinant clones can be easily distinguished from the nonrecombinants based on the color of the colony. It is a single-step process, unlike replica plating. The chromogenic technique was introduced in pUC vectors for the first time and the ease in identification of recombinant clones in, this technique has extended it to a wide variety of vectors (pUC series, pBluescript, pGem, and their derivatives). This method is popularly called a blue-white screening technique. The β-galactosidase enzyme is encoded by *E. coli* and converts lactose into glucose and galactose. The active β-galactosidase enzyme can hydrolyze X-Gal (an analog for lactose) into insoluble deep blue-colored products. This phenomenon was first discovered by Julian Davies in 1967 to detect the colonies carrying the active β-galactosidase enzyme) [3].

In this screening method, the plasmid vectors synthesize a short segment of the first 146 amino acids of the β-galactosidase enzyme encoded by the lacZ gene. The *E. coli* host cells used to carry such vectors are mutant for lacZ gene (lacZΔM15 deletion mutation), which encodes for the carboxyl end of the β-galactosidase enzyme. Independently the vector and the host cannot synthesize a functional enzyme, but when the host carries such vectors, due to α-complementation, the functional β-galactosidase enzymes are synthesized. This phenomenon with certain modifications is used in vector systems to identify the recombinant clones effectively.

As shown in Fig. 7.3, multiple cloning sites (MCS) can be found within the LacZ gene in a plasmid vector (ex: pUC). The gene of interest can be inserted within the vector by cleaving the MCS by a restriction enzyme. The ligated product is transformed into the host *E. coli* system and cultured on a plate containing antibiotic (for selecting transformants), Isopropyl β-D-1-thiogalactopyranoside (IPTG: which acts as an inducer for the LacZ promoter), and X-gal (as a substrate). If the *E. coli* cells take up the recombinant vectors (plasmids with the gene of interest), the colonies remain white. The gene of interest will be interrupting the expression of the Lac Z gene, which results in lack of alpha complementation, and, thus X-gal is not hydrolyzed so the colonies remain white. On the contrary, if the colonies carry only the vector backbone. The Lac Z gene is intact and therefore the cells are capable of synthesizing the functional β-galactosidase enzyme due to α-complementation, which hydrolyzes the X-gal into 5-bromo-4-chloro-indoxyl, which spontaneously dimerizes to produce an insoluble blue pigment 5,5′-dibromo-4,4′-dichloro-indigo, giving blue color to the colonies. So in general in blue-white colony assay, blue-colored colonies are nonrecombinants and white colonies are nonrecombinants [4].

Box 7.1

β-galactosidase gene can also be used as a reporter gene to identify the recombinant clones successfully. Reporter genes are the one that synthesizes a product that can be easily recognized and measured. Similar to the β-galactosidase gene, other reporter genes like fluorescent genes (GFP, RFP) and enzymes like luciferase. A chimeric construct is produced in the reporter gene assay system by enclosing a regulatory element before the reporter gene. If the regulatory elements are functional, the reporter gene gets expressed and can be measured. This technique is commonly used in gene expression and regulation studies. Recently automation, electrochemical sensors and biosensors has allowed rapid and precise screening of recombinants and their products in cost effective ways [5]. Due to microfluids, it is now possible to work in microliter to picoliter range or even can be extended up to single-cell analysis. Microbioreactors is a yet another example of miniaturization of traditional in vitro culturing and fermentation techniques which significantly cuts down the volume of sample necessary for screening. Even though a variety of fluorescence reporter are typically used in screening, the evolution of biosensors are changing the magnitude of screening and detection thresholds. The DNA or protein part of the biosensor identifies the target while reporter amplifies it to measurable electrochemical signal, for example hybrid transcription factors biosensors or the engineered fluorescent protein biosensors.

Advantages

1. It is a quick and more straightforward method to screen the transformed colonies.
2. It helps in differentiating between bacteria and phage particles which may contain the gene of interest in cloned vectors and refuse empty vectors,
3. The method is visual-based, which helps segregate blue and white colonies and no need for automation or highly advanced analytical techniques.
4. It is a substrate-dependent method (X-gal) which makes it more convenient.

Limitations

1. White colonies may or may not contain the gene of interest.
2. Even if a gene of interest is present, lacZ may not be functional and may not produce β-galactosidase, which may not convert X-gal to blue substance and resulting white colonies may not be recombinants.
3. There is also a possibility that blue colonies may contain the gene of interest, but because of visual appearance, the chance of rejecting such colonies will be high.
4. There is a possibility of insertion of foreign DNA in the MCS region, which will alter the reading frame of lacZ that can develop false-positive white colonies.
5. Sometimes complete transformed colonies may also produce blue colony leading to confusing results.
6. X-gal is expensive and unstable.

7.3.2 Positive Selection

In positive selection, only the clones which receive the foreign DNA or gene of interest survive. It avoids the other background noise of nonrecombinant clones, mainly by limiting the growth of transformants with self-ligated and uncut vectors. In this technique, the vector gene that encodes for some lethal protein is turned off either by replacing or inactivating the corresponding gene. Generally, the latter one is preferred because of its simplicity and accurate result.

A wide variety of functional genes encoded by the vector and whose products are lethal to host cells can be applied to select only recombinant clones from the mixture of transformants. One such example is the coupled cell division (CCDB) gene, which encodes a protein that hampers the activity of topoisomerase II (Gyrase) activity in prokaryotes. Without the proper functioning of the enzyme gyrase, it results in irreparable DNA damage, thereby killing the cells. In a vector, the CCDB gene is tagged along with the LacZ reporter gene which is under the control of the lac promoter. With this arrangement in a vector, the CCDB gene can be activated by inducing the LacZ promoter using IPTG as an inducer. Similarly, the gene can be inactivated by inserting the gene of interest in the MCS which is located within the

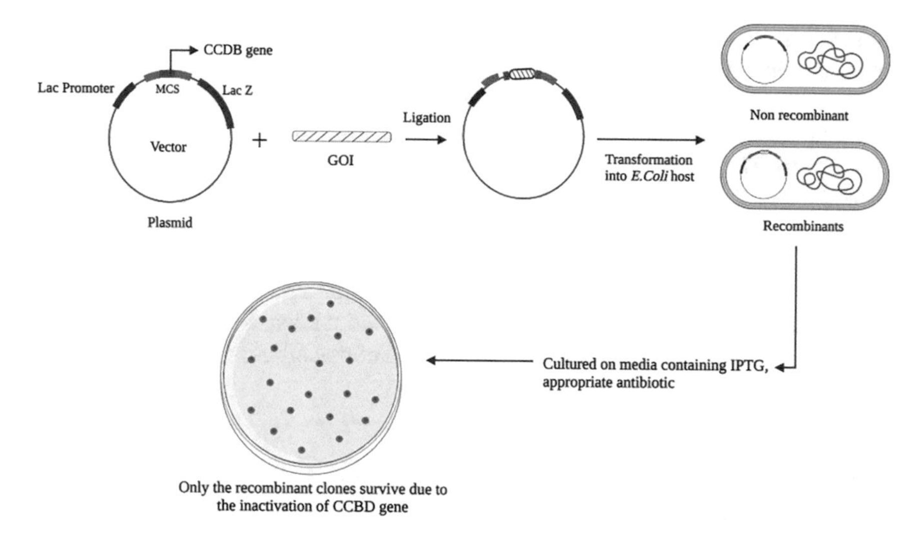

Fig. 7.5 Positive selection of recombinants by inactivating the CCDB gene encoded by the vector

CCDB gene. This restores the activity of the enzyme gyrase. Consequently, the transformants which receive the recombinant vectors survive and the ones which harbor nonrecombinants (Fig. 7.5) (where the CCDB gene is intact) do not survive [1].

Positive selection vectors hold a more significant advantage over the average cloning vector in selecting the transformants with the gene of interest, as the plate contains only recombinant cells after transformation.

Advantages

1. Vectors impart an advantage of selecting only the recombinants by reducing the background noise created because of nonrecombinant clones.

Limitations

1. Low availability of specific host: The few positive selection techniques requires special host strains (like in the case of yeast the auxotrophic mutants).
2. In some techniques minimal media are used where the colonies take more time to grow.

7.3.3 Selection Based on Genome Size

In this technique, the selection of recombinant clones is done based on their genome size. This method is mainly followed in λ phage particles, where there is a direct link between the genome size and the packaging of the genome into a protein coat to produce mature phage particles. Normally, when the phage particles infect the host cell genome, it utilizes its machinery to synthesize the viral genome and the coat protein separately. Later on, the viral particles are matured by packing their genome into the protein coat. During packing only, the genome size between 37 kb to 52 kb gets packed and the genome whose size does not fall within this range is not packed. The phage vectors are constructed by deleting a large part of the nonessential phage genome which reduces its size, thereby restricting itself from packaging inside the protein coat. These modified vectors are then incorporated with foreign DNA segments making its genome size fall between the packaging range. Thus only the recombinant phage particles are packed and mature phage particles are produced thereby forming plaques on the bacterial lawn. So automatically only the recombinant phage particles survive over the nonrecombinants.

7.4 Screening of Recombinant Clones for the Specific Gene of Interest

The question of a specific gene arises when there is a need to identify the clones with a target gene/ specific DNA segment among the large pool of recombinant clones carrying unique DNA segment/gene in them. One such example is a screening of genomic or cDNA libraries for the desired clone, wherein, the genomic library is a collection of clones and should represent every segment of an organism's genome at least once. The genome size of prokaryotes and lower eukaryotes is smaller in nature and hence a smaller number of recombinant clones is required to represent the entire library. Owing to the larger genome size of plants and animals necessitates a high number of clones to represent a complete library which makes the identification of the target clones a complicated and time-consuming process.

To reduce the screening effort in the case of eukaryotes, cDNA library can be constructed which is cellular specific. This represents only the genes that are expressed in the particular cell type. The cDNA library is constructed by converting mRNA into complementary DNA (cDNA) by reverse transcriptase enzyme. Thus the cDNA library represents fewer clones making the identification process relatively easy compared to the genomic library. However, the cDNA library also consists of large number of clones making it challenging to identify the clones with the required DNA segment. Hence, it is imperative to use a sensitive, precise, and high throughput screening strategy to identify the target clones [6].

Many such screening techniques are available, where the translated products of the gene of interest can be used to identify the desired clones. A rapid and easy method is to make use of DNA/RNA probes, which will hybridize with the specific gene of interest because of the presence of a complementary sequence. In this section, we will see how to screen the library for target clones by utilizing translated products of the gene using DNA/RNA probes that complement the DNA segment of interest.

7.4.1 Nucleic Acid Hybridization Technique

The success of this technique mainly relies on the accuracy with which the probe binds with its complementary target sequence. It mainly depends on the percent homology of probes with that of the target sequence. If the probe is designed using the sequence of the same organism, 100% homology is expected between the probes (Homologous probes) and the target sequence. Using these homologous probes the target clones can be identified at a higher rate. On the contrary if the probes are designed from the sequence of the related organism, the homology between the probe and target DNA reduces (heterologous probes). Nevertheless, these heterologous probes are very helpful in identifying similar genes from different sources but the success rate reduces. In such a situation to avoid the binding of probes to the nonspecific target region and to increase the specificity of hybridization, the hybridization conditions can be more stringent. Conversely it can be less stringent in the case of homologous probes. The reduction of stringency in case of heterologous probes leads to an increase in the hybridization, by tolerating a certain level of mismatching which helps in the identification of related sequences from a different source.

Usually the DNA/RNA probes are single-stranded molecules of around 50 to 500 bp length and are labeled with either radioisotopic or non-radioisotopic molecules to detect the molecule after hybridization. Even though radioactive labeling is highly sensitive, it is hazardous to the environment. In recent years nonradioactively labeled probes have been in the limelight, wherein the enzymes are used for labeling and the hybridized product can be detected based on the change in either color or by the emission of illuminance or fluorescence due to the activity of an enzyme on the substrate molecule. The availability of the aforesaid probes helps in the identification of any target sequence from any library by increasing the efficiency of the hybridization process. If the DNA/RNA probes are not being available, the knowledge of the protein synthesized by the target gene and its amino acid sequence can be used to synthesize the short stretch of DNA/RNA oligonucleotide chemically and used as probes for detecting the target gene in a pool of recombinant clones.

7.4.2 Colony or Plaque Hybridization Technique

This is one of the membrane-based rapid techniques used to identify the recombinant clones with the target DNA segment from a clone bank. The clone bank consists of a large number of clones with various hybrid plasmids harboring different DNA segments. Here either DNA or RNA hybridization probes specific to the target gene are used.

In this technique, replica plating (discussed in Sect. 7.3.1.1) of the master plate is done on a nitrocellulose membrane, which acts as a blotting medium where exactly the hybridization between the DNA sequence and the probe takes place. The

DNA/RNA probes are radioactively labeled. It can be detected by autoradiography. If it is nonradioactively labeled, it is detected by either change in the color or emission of luminescence or fluorescence. The position of the colony hybridized is traced back to the master plate. Thus the recombinant clone with the target sequence is recovered.

In colony hybridization, the clones are carefully transferred to a nitrocellulose membrane from the master plate. The bacterial cells are lysed by treating the membrane with Sodium dodecyl sulfate and proteinase k. The content of the cells along with the DNA will get released which is then treated with alkali to separate the two strands of the DNA by breaking the hydrogen bond. The filter papers are then baked to 80°C. At this high-temperature DNA (ssDNA) binds to the membrane with the help of the sugar-phosphate backbone exposing the bases. At this point, the membrane is treated with the labeled DNA or RNA probes. The probes get hybridized with the complementary sequence at proper hybridization conditions. After hybridization, the excess probes that are not hybridized are washed off to reduce the background noise. The membrane is exposed to autoradiography which detects the colonies that carry the target DNA sequence which can be traced back in the master plate (Fig. 7.6).

The phage vectors are commonly used to build gene banks rather than plasmids as their insert DNA carrying capacity is much larger compared to plasmids. Here the plaques carrying the different foreign DNA segments are grown on the lawn of bacteria. Later on, the nitrocellulose membrane is used to get the replica of the master plate. The plaques carrying the target segment can be detected by following steps similar to colony hybridization.

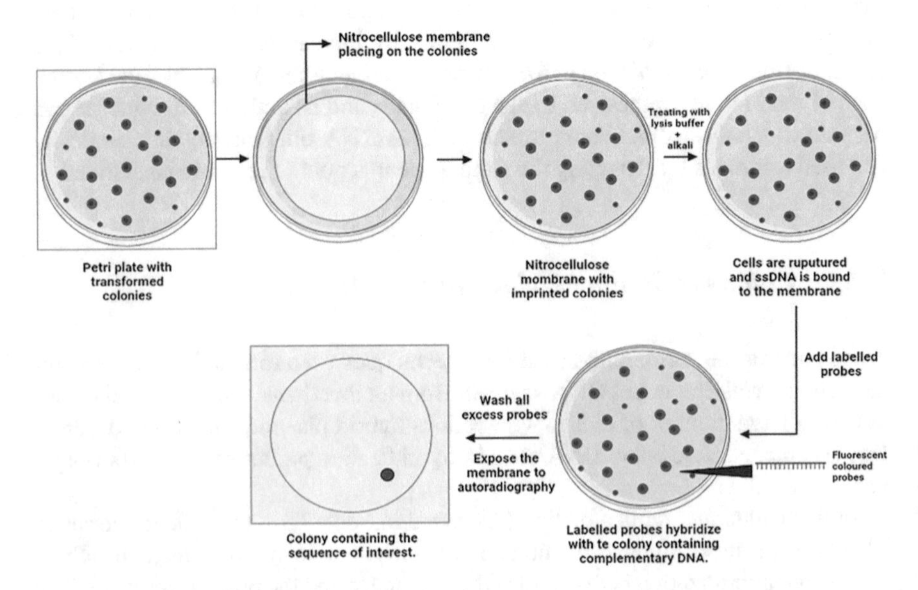

Fig. 7.6 Colony hybridization to detect the recombinant clone carrying the target sequence

Advantages

1. It has less background probe hybridization for colony screening which gives an advantage of producing clean and distinct signals.
2. Requires less DNA to transfer from host cells to the membrane by disrupting the bacterial colonies on the membrane itself.
3. Multiple screening is possible because the same colonies can be used several times.
4. It also enables the screening of a large number of recombinants homologous to a probe in one single experiment.

Limitations

1. As the method is based on probe hybridization. The probes should be designed with high precision and accuracy.

7.4.3 Immunological Assay to Select the Recombinants

This is the technique wherein the recombinant clones with the target gene can be identified based on the translational product of the gene itself. In a situation where information on the DNA sequence is not available. Designing DNA or RNA probes is difficult. Nevertheless, suppose the protein synthesized by the gene is well studied, and specific antibodies are available. These antibodies can be used to hybridize with the protein encoded by the target gene, thereby identifying the target clone. (Box: Antibodies are immunoglobulin protein fraction produced by animals as an immunogenic reaction in response to antigens (proteins). They bind to the antigen and help in its degradation. There are mainly two types of antibodies used for hybridization purposes: polyclonal and monoclonal antibodies (synthesized by injecting the antigens in the mouse body, and antibodies are collected from the serum as an immunoglobulin portion). The Polyclonal antibodies act against all the determinants of the antigen, whereas monoclonal antibodies act against the specific antigen determinant) [7].

As this method completely relies on the protein antibody interaction expression of the cloned gene is important. Hence the library should be constructed using expression vectors. In the case of eukaryotes, it is more appropriate to use cDNA libraries to identify the target clone while in case of prokaryotes, the genomic library itself can be used owing to the fact that a minor portion of the genome of the prokaryotes harbors noncoding sequence (introns and repetitive sequences) and most of the genome codes for proteins [8].

This method is similar to colony hybridization, however, instead of using DNA probes here, the antibodies are used which will go and bind to the specific antigen determinant synthesized by the clones with the target gene. Similar to colony hybridization here also the replica of the master plate is made on the nitrocellulose membrane (Fig. 7.7). The membrane is treated with chemicals and enzymes to

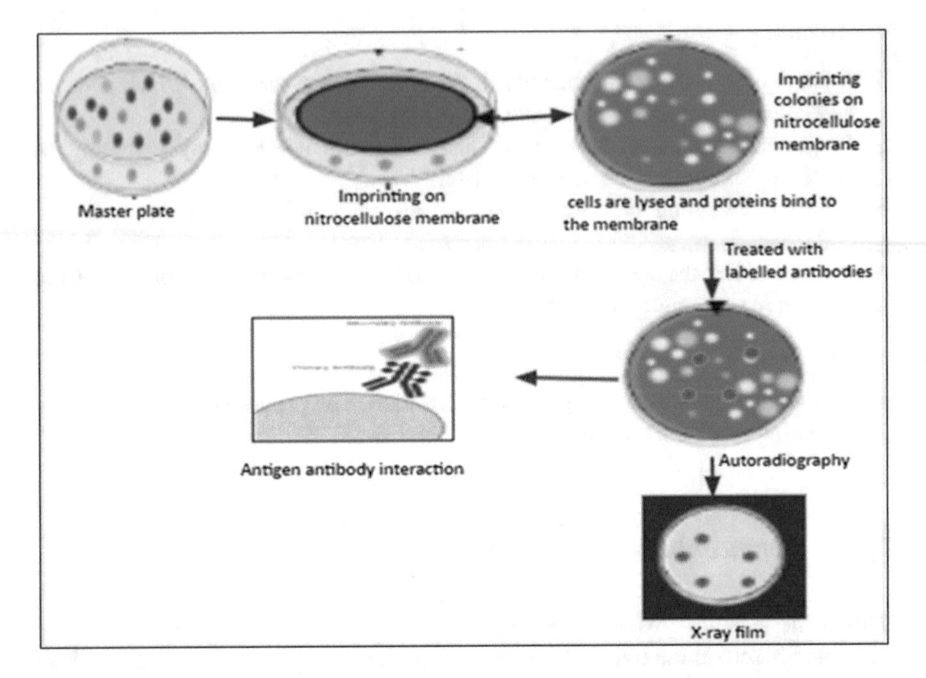

Fig. 7.7 Selection of target recombinants based on translational product of the transgene

disrupt the cell membrane and the contents are released. Later on, the hybridization can be followed in two ways. The first method, the lysed colonies are treated with antibodies labeled with either radioactive (I125 isotope) or nonradioactive substances. The target colony can be detected by exposing the membrane to autoradiography or chemiluminescence/chemifluorescence/colorimetric technique. In the second method, the lysed cells are treated with antibodies called primary antibodies. After that, the secondary antibodies interacting with the primary antibodies are used and labeled with either radioactive or nonradioactive substances. The use of primary and secondary antibody sandwich method will enhance the detection signal, and the target clones can be identified with higher accuracy [8, 9].

Advantages

1. It is based on colony selection, and it is simple and easy to pick up the recombinant colonies.

Limitations

1. To develop a specific antibody for each protein is a costly and time-consuming process.
2. For large-scale production of specific antibodies, animals need to be injected for the production of foreign protein or peptide, which is laborious and uneconomical.
3. This procedure is suitable only when a large amount of protein is produced, but while screening for recombinants, the production of a large amount of protein will have a practical limitation.

7.4.4 PCR-Based Technique

PCR is the technique that is more commonly used in Molecular Biology wherein the primers specific to the target segment of the DNA (gDNA or cDNA) are used for amplification of the DNA fragment in millions of copies. This technique can be extended to screen genomic or cDNA libraries to detect the target clones.

This method can easily replace the probe hybridization technique as it is the quickest method to screen the recombinants. However, this method is also associated with some limitations, like the availability of sequence formation to design the suitable primers. This can be used to screen clone libraries with certain modifications in the regular PCR, as suggested by Takumi & Lodish, 1994. In the modified PCR, the colonies from the master plate are maintained as pooled colonies in a multiwell plate. Next, the cells are lysed, and this pooled mixture of DNA is used as a template to carry out PCR using a set of target DNA-specific primers. Finally, set PCR using the same set of primers for individual colonies from the master plate, which are part of the positive pool. Thereby the recombinant target clones can be detected more accurately and rapidly [2].

Advantages

1. Colony PCR is the quickest method for screening recombinant clones. The primers specific to the gene of interest or a combination of one primer from the vector and other from the gene can be used to confirm the gene of interest and to know the direction of the inserted gene.

Limitations

1. It is based on specific primer designing, the primers need to be designed with high precision and accuracy.

7.4.5 Functional Complementation of the Gene Product

This method can be applied when the information about the gene is not available except for its function. In such conditions, complementing the missed function in the host cell by inserting the vector carrying the gene for the function is used. Thus by functional complementation, the recombinant clone carrying the suitable gene can be recognized. For example, the shaker-2 homozygous mutant mouse was deaf in nature. The recombinant clones of the library developed from the wild mouse (using BAC vectors) were transferred to the shaker-2 mutant egg cell and progenies were recovered. The transgenic mouse which received the recombinant clone carrying the gene for correcting the deafness exhibited correction in the shaker-2 phenotype and their hearing capacity was restored (Fig. 7.8). This helps in the identification of the recombinant clone carrying the shaker-2 gene more precisely because if there is no functional complementation, the wild trait will not be restored [10].

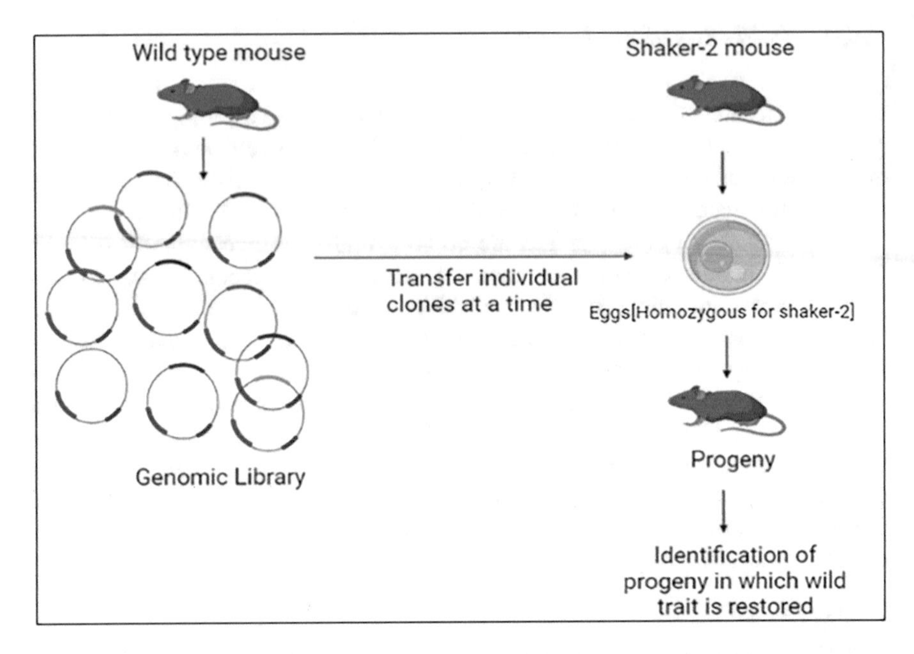

Fig. 7.8 Selection of the desired clone based on functional complementation of the gene

Advantages

1. The method is very accurate in selecting the target recombinant as the recombinant clones which do not carry the suitable gene do not express the wild trait due to the absence of complementation.

Limitations

1. The success of the technique completely depends on the availability of mutant expression hosts.
2. It is time-consuming because in the case of eukaryotes (plants and animals) it requires the development of progenies to know the restoration of the function of the gene.
3. An expression library is required to perform the technique.

High Throughput Screening and Selection Techniques

The limitations of traditional selection and screening techniques can be overcome by high-throughput screening methods that can obtain recombinant products. These methods have the advantage of automation [11, 12]. A few high throughput screening techniques have been mentioned in Table 7.1.

Table 7.1 List most commonly used high throughput techniques to screen the recombinant clones

Sl. No	Screening method	Features	Advantages	Disadvantages
1.	Microtiter plate	It is an enzyme-based technique, wherein changes in the color, pH, fluorescence, cell growth, etc., are observed to identify the desired recombinants among the transformed colonies.	1. Plate techniques like ELISA, which are small test tubes assembled on a single plate, can be loaded with multiple experimental materials on a single plate yet separate each experiment in every well. 2. The plates can also be stored for further analysis and future use in refrigerated conditions.	1. It is a Laborious process
2	Digital imaging (DI)	It is the screening of solid-phase through single-pixel imaging spectroscopy.	1. It can be utilized where substrates are costly or problematic and can be screened based on colorimetric assays and highly sensitive, leading to precise results.	1. It applies only to colorimetric assays and is highly specific.
3	Fluorescence-activated cell sorting (FACS)	It is a fluorescent-based individual cell signaling technique that sorts cells into two or more containers up to 30,000 cells/s.	1. Extremely high throughput, sensitivity, and reduced time consumption. Individual cells can be sorted out with less time.	1. The enzyme activity depends on the fluorescence of the target protein. 2. If there is a change in fluorescence or intensity of light, it can lead to ambiguous selection and screening of recombinants.

(continued)

Table 7.1 (continued)

Sl. No	Screening method	Features	Advantages	Disadvantages
4	Cell surface display	The enzymes encoded by DNA within the cell fused with anchoring motifs are expressed on the cell surface and bind to the substrate.	1. Cell surface display can be combined with FACS and FRET techniques resulting in a huge number of cell screenings at one time, 2. Cell lysis is not required.	1. Proteins expressed are fused and may not be ideal when looking forward to a single protein.
5	In vitro compartmentalization (IVTC)	IVTC uses compartments like W/O (water-oil emulsion) or W/O/W (water-in-oil-in-water) emulsions to isolate individual DNA molecules, leading to the formation of cell-free protein synthesis and enzyme reaction.	1. As there is no transformation, library size can be more, avoiding transformation efficiency of host cells. 2. It can be combined with FACS and screening efficiency is very high and sensitive.	1. As there is no posttranslational modification, it is not suitable for the enzymes which are interphase between transcription and translational level.
6	Resonance energy transfer (RET)	It is the mechanism of energy transfer between two chromophores or fluorophores. The donor will transfer energy to the acceptor through nonradioactivity. GFP, CFP, YFP, and RFP chromophores are widely used in RET based on protein–protein interactions.	1. The highly efficient and high throughput screening method.	1. As the enzyme activity is associated with two fluorophores makes it a significant limitation. 2. The technique is nonradioactive.

7.5 Analysis of Recombinant Clones

After identifying the recombinant clone, it is necessary to go for a detailed characterization of the transgene for further applications in commercial use in advanced research activities. There are different approaches to analyzing the transgene, but the selection of the method mainly depends on the availability of information about the transgene or based on the ultimate motive of the experiment.

7.5.1 Restriction Analysis

The genomic and cDNA libraries are constructed using cosmid or bacterial artificial chromosomes (BAC) and their insert carrying capacity is much larger. As a result, it may carry multiple genes in a single clone or only part of the gene may be present in some cases. So after the identification of the desired clone, characterization of the gene is essential. One of the approaches is to cleave the transgenes using various restriction enzymes to develop a restriction map for the gene. With the availability of restriction maps, a particular segment of DNA can be selected and used for various applications. Initially, the transgene is cut with a variety of restriction enzymes and among those, which cut the transgene into an average of 3 to 4 fragments are selected. Using these enzymes, single or multiple digestion is done to get a detailed restriction map of the transgene.

7.5.2 Hybridization Technique

Developing a restriction map for the transgene is not sufficient. It is just the first step in getting the right segment of the DNA present within the transgene. The length of the transgene within the vector can vary from 25 kb (λ Phage vector) to 300 kb (BAC vector) in a library. Every time it is not economical to go for sequencing such a long segment of DNA, to find out the sequence of interest. Instead, a restriction map of the transgene is developed to identify the exact gene sequence by using labeled DNA/RNA probes. These probes will go and bind to the target fragment carrying the gene sequence and are detected using appropriate detection techniques. To achieve this, the restriction of digested DNA fragments of the insert is transferred to an agarose gel, where the fragments are separated based on their sizes. These separated fragments are then transferred to the nitrocellulose membrane called blotting (Fig. 7.9). These nitrocellulose membranes are then treated with various chemicals to separate the two strands of the DNA to ssDNA and to fix them on the membrane. Finally, these are treated with labeled probes, which will hybridize with the complementary ssDNA sequence present on the membrane. Unhybridized probes are washed off. The membranes are exposed to appropriate detection techniques based on the type of molecule used for the labeling of probes. If DNA probes are used for hybridization then it is called southern blotting and if RNA probes are used it's called northern blotting.

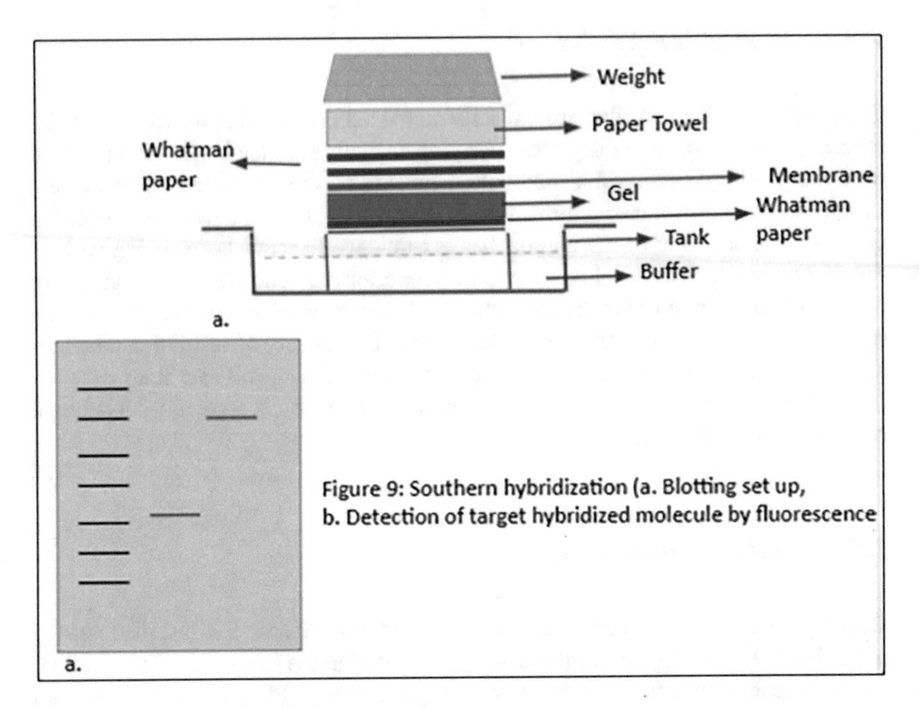

Fig. 7.9 Southern Hybridization (**a**) Blotting setup; (**b**) Detection of target hybridized molecule by fluorescence

7.5.3 DNA Sequencing

All the above-discussed approaches like restriction analysis and hybridization techniques help ascertain the presence of part of the DNA sequence of interest within the transgene in the identified clones/fragment. However, to know the exact coding sequences, presence of regulatory elements, and any other features sequencing of the identified sequence is a must. Usually, the di-deoxy sequencing method is followed to find out the order of bases arranged in a target sequence. Sometimes only part of the gene is present in one clone and the remaining part in another clone. The two clones are sequenced independently and both the sequences are aligned and the region of overlapping helps in identifying the adjoining sequences. Hence sequencing is the most accurate way to cross-check the recombinant clone and verify the sequence of interest within the transgene.

Self Assessment

Q1. Differentiate between selection and screening in the context of transformants and recombinants.

A1. Selection refers to picking out only the transformed cells or cells containing vectors (which may contain genes or may not) from the non-transformants (cells without any vector) after transforming the host cells with the desirable vector. It is achieved by exerting pressure under certain defined conditions. Whereas in screening the recombinant clones (transformants harboring vectors with foreign DNA segment) are separated from the nonrecombinant clones (transformants carrying only vectors without any foreign DNA segment in them) among the pool of transformants.

Q2. Which one is the quickest method to select the recombinants? Why?

A2. Selection of recombinant clones by polymerase chain reaction (PCR) is the quickest method of screening as we get result in 3 hrs with the use of forward and reverse primers specific to the foreign DNA segment. Even a large number of clones can be screened at a time its more quicker compared to other selection strategies.

References

1. Young-Jun C, Tsung-Tsan W, Lee BH. Positive selection vectors. Crit Rev Biotechnol. 2002;22(3):225–44.
2. Tsuji T, Onimaru M, Yanagawa H. Random multi-recombinant PCR for the construction of combinatorial protein libraries. Nucleic Acids Res. 2001;29(20):e97.
3. Guo Y, Hui CY, Liu L, Zheng HQ, Wu HM. Improved monitoring of low-level transcription in Escherichia coli by a β-galactosidase α-complementation system. Front Microbiol. 2019;10:1454.
4. Green MR, Sambrook J. Screening bacterial colonies using X-Gal and IPTG: α-Complementation. Cold Spring Harbor Protocols. 2019;2019(12). pdb-rot101329
5. Zeng W, Guo L, Xu S, Chen J, Zhou J. High-throughput screening technology in industrial biotechnology. Trends Biotechnol. 2020;38(8):888–906.
6. Bazan J, Całkosiński I, Gamian A. Phage display—a powerful technique for immunotherapy: 1. Introduction and potential of therapeutic applications. Hum Vaccin Immunother. 2012;8(12):1817–28.
7. Xiao H, Bao Z, Zhao H. High throughput screening and selection methods for directed enzyme evolution. Ind Eng Chem Res. 2015;54(16):4011–20.
8. Broome S, Gilbert W. Immunological screening method to detect specific translation products. Proc Natl Acad Sci. 1978;75(6):2746–9.
9. Lo RY, Cameron LA. A simple immunological detection method for the direct screening of genes from clone banks. Biochem Cell Biol. 1986;64(1):73–6.
10. Probst FJ, Fridell RA, Raphael Y, Saunders TL, Wang A, Liang Y, Morell RJ, Touchman JW, Lyons RH, Noben-Trauth K, Friedman TB. Correction of deafness in shaker-2 mice by an unconventional myosin in a BAC transgene. Science. 1998;280(5368):1444–7.
11. Shaw-Bruha CM, Lamb KA. Ion pair-reversed phase HPLC approach facilitates subcloning of PCR products and screening of recombinant colonies. BioTechniques. 2000;28(4):794–7.
12. Choe J, Guo HH, van den Engh G. A dual-fluorescence reporter system for high-throughput clone characterization and selection by cell sorting. Nucleic Acids Res. 2005;33(5):e49.

Chapter 8
Recent Trends and Advances

Shruti Desai, Nayana Patil, and Aruna Sivaram

8.1 Introduction

The monumental discovery of the double helix in the 1950s was heavily dependent on the ground work of researchers, way back in the 1800s. It started with the discovery of the genetic traits, nuclei, confirmation of Mendel's theories in human diseases further identifying DNA as the transforming principle and the discovery that DNA composition is species specific. In 1952 Rosalind Franklin crystallized the DNA fibers leading to the discovery of the invisible genetic material. From then, the trajectory of understanding the complexity of DNA has been only steadily increasing.

Manipulation of DNA or RNA developed in the early 1970s has established the field of genome editing, showing its powerful use in the areas of basic research, agriculture, medicine, industrial products, etc. Damaging agents outside any cell or even internal metabolic byproducts can lead to alterations or damages in the genome. The damage caused due to these assaults cause breaks in the DNA that could result in single-stranded breaks (SSB) or double-stranded breaks (DSB). These breaks are lethal to the cells. Hence, the cells rely on pathways [1] like non-homologous end joining (NHEJ) and homology directed repair (HDR) for its mending the breaks as shown in Fig. 8.1.

S. Desai
Department of Pathology, Yale University School of Medicine, New Haven, CT, USA
e-mail: shruti.desai@yale.edu

N. Patil · A. Sivaram (✉)
School of Bioengineering Sciences & Research, MIT ADT University,
Pune, Maharashtra, India
e-mail: nayana.patil@mituniversity.edu.in; aruna.sivaram@mituniversity.edu.in

N. Patil, A. Sivaram, *A Complete Guide to Gene Cloning: From Basic to Advanced*,
Techniques in Life Science and Biomedicine for the Non-Expert,
https://doi.org/10.1007/978-3-030-96851-9_8

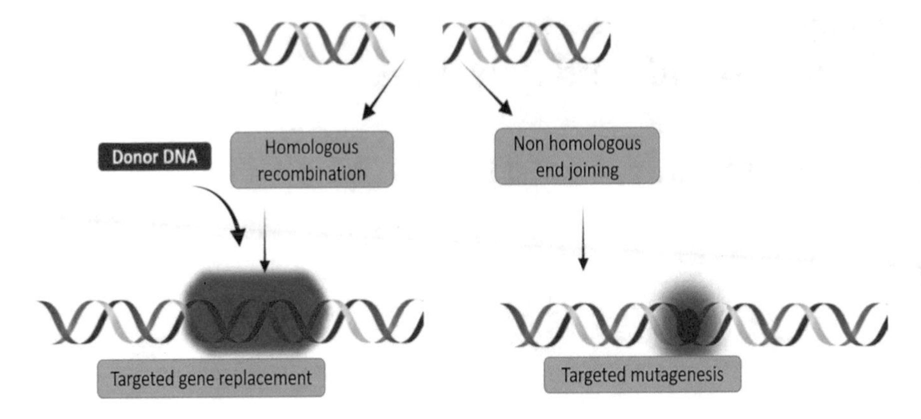

Fig. 8.1 The different mechanisms of DNA repair after double-stranded break repair

The following examples show how different types of DNA repair processes are utilized to obtain the targeted alterations in the cells/ organisms. These processes form the foundation of genome editing.

- *Targeted gene mutation/knockout*: This method utilizes the error prone property of NHEJ at the target site and introduces small indels. Targeting the coding region of a gene often results in indels, causing frameshifts.
- *Targeted gene deletion*: It includes deleting small or large segments of DNA by introducing two DSBs. This method does have a few limitations because it could eliminate an entire genomic element.
- *Targeted gene correction*: In this method gene editing is achieved by activating HDR with an external donor DNA, which is used as a repair template. Plasmids, single stranded oligonucleotides (ssODNs), lenti/adenovirus can serve donor DNA templates for HDR.
- *Targeted gene insertions*: Although viral vectors are efficient delivery agents, these viruses are not reliable for site-specific integration and there are concerns of accidental mutagenesis. Hence targeted gene insertions into specific genomic sites alleviate the hazards of insertional mutagenesis. Targeted transgene insertion can be achieved by nuclease induced DSBs that create compatible overhangs on the donor DNA, facilitating NHEJ-mediated ligation of the required DNA fragment into the target locus.

The success of genome editing is fundamentally based on accurately locating the correct piece of DNA that needs to be fixed and altering it at specific sequences. The significance of gene alteration that depends on the DNA repair mechanism is that it is applicable to a majority of cell types or organisms as it relies on the endogenous processes. In this chapter we will discuss a few major platforms used to introduce site-specific DSBs such as zinc finger nucleases (ZFN), transcription activator-like effector (TALE)-nuclease (TALEN), CRISPR/Cas system, CRE-LOX model, and RNA interference (RNAi).

8.2 Zinc Finger Nucleases (ZFN)

One of the primary and most explored approaches of targeted gene editing is zinc finger nuclease (ZFN). The ZFN are artificial endonucleases that have the exceptional ability to recognize unique target sequences in the eukaryotic genome, hence are used in targeted genome editing. ZFN are also used in regulating the expression of specific genes which involves tagging of the endonuclease dimer with activation or repression domains to turn a gene on or off respectively [2].

The ZFN has two domains—a eukaryotic transcription factor with zinc finger motifs and a *Fok*I restriction endonuclease. The Zinc finger plays the pivotal role of identifying and locating the target sequence while the nucleases create DSBs which are subsequently repaired by the DNA repair system resulting in site-specific modification. Architecturally the most important part of ZFN is that the DNA cleavage site requires nuclease dimerization. The Zinc finger proteins and wild-type *Fok*1 are inactive as individual entities and can function only when the *Fok*1 monomers interact through cleavage domains [3]. To target a particular site in the genome, it should be recognized by two ZFN subunits on opposite strands in a tail-to-tail orientation and these "half-sites" should be separated by a 5–7-bp spacer sequence. Thus, ZFN is successful when the sequence located by zinc finger protein is cleaved after the dimerization of nucleases.

8.2.1 DNA-Binding Domain

The zinc finger proteins form the DNA-binding domain of the ZFN. The eukaryotic zinc finger proteins are small proteins consisting of 28–30 amino acids forming β-hairpin (antiparallel β-sheet consisting of two β-strands) followed by an α-helix, which form ββα conformation. These proteins are classified as C2H2-type due to specifically conserved cysteine and histidine involved in zinc binding. The 3D folding of the protein brings the pair of histidine in α-helix at the carboxy terminal in close proximity to the two cysteines at the end of the β-strand enabling them to bind to Zn^{2+} to form a tetrahedral structure. The interlocking between an α-helix and two antiparallel β-strands due to Zn^{2+} provides thermal and conformational stability and stabilizes the fold resulting in a typical finger-like appearance of the protein. The side chain of conserved amino acids also permits binding of zinc finger protein to the major groove of the DNA and reinforce the specific binding and stabilization of the complex.

ZFN has a modular assembly of several zinc fingers connected through highly conserved TGEKP linkers. A typical ZFN consists of three to six different zinc fingers repeats, each recognizing a three-base pair DNA sequence thus locating a target site ranging up to 18 base pairs in length (Fig. 8.2).

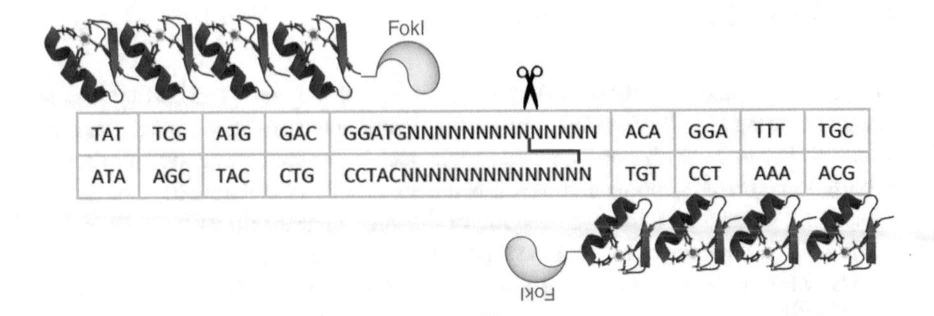

Fig. 8.2 Basic structure of a zinc finger nuclease (ZFN). A single zinc finger motif recognizes a triplet in DNA, three or four such motifs are linked in an array to form left and right monomers; the tandem array is linked to *Fok*I nuclease domain. The ZFN binds predetermined targets on opposite strands bringing two *Fok*I units together; the dimerization induces a double strand break on its target

8.2.2 DNA-Cleavage Domain

The *Fok*1 enzyme forms the DNA cleavage domain of ZFN. A type IIS restriction enzyme, *Fok*1 is an integral part of ZFN. The *Fok*1 cleavage domain has no defined sequence specificity and is solely guided by zinc finger protein. The C-terminal Cys_2-His_2 domain of zinc finger protein is glued to the cleavage domain using 6 amino acid linkers (TGAAAR). The binding of a pair of ZFN to non-palindromic DNA in an inverted orientation allows dimerization of two halves of the cleavage domain constructing a functional *Fok*1 thereby making a DSB. By this way, the specificity of the system improves and toxicity is reduced.

*Fok*1 is a type IIs restriction endonuclease from *Flavobacterium okeanokoites.* The enzyme is 65.4 kDa protein, comprising a 41 kDa N-terminal fragment with DNA-recognition function and 25-kDa C-terminal fragment with DNA cleavage function. The recognition domain is made of three smaller subdomains (D1, D2, and D3) and the cleave domain has two loops P1 and P2. The catalytic loops are sequestered by DNA-recognition sequence and swing free only when *Fok*1 binds to DNA substrate. The two individual monomers at the 3′ end of the recognition sequence collide and in presence of magnesium ion align its reaction center leading to activation of the catalytic domain thus generating a double-strand break (DSB). The active restriction enzymes can now recognize 5′-GGATG-3′ sequence and cut the DNA strand 9 base pairs downstream of this site and 13 base pairs upstream of the recognition site on the complementary strand. Hence, the recognition site is represented as $GGATG(N)_{9/13}$.

8.2.3 Applications of ZFNs

The discovery of ZFN paved the way for all advancements that we see in the field of genome editing. The applications of ZFN can be seen in almost all walks of life. For example, in the field of agriculture [4], the endochitinase gene in tobacco was targeted using ZFNs and donor DNA encoding a herbicide resistant marker was introduced. Enhancement of seed oil profile in soybean by adding *FAD2* (*fatty acid desaturase*) gene into the soybean genome and imidazolinone herbicide resistant wheat genotype was confirmed by single gene mutant in *acetohydroxyacid synthase* using ZFN technique. ZFNs brought in a lot of promise for efficient creation of genetically modified animal models especially rats and mice [5]. In 2009 ZFNs were used to target the GFP sequence of transgenic rats and knock out the target gene without any off-target effects. Similarly, a SCID knockout rat model was generated by Mashimo et al. Stable transgenesis of an inducible ZFN expression for gene disruption has shown to be successful in Plant model *A. thaliana* where mRNA microinjections were not an option.

Due to its efficiency and less off-target effects ZFNs have been specifically used in the glioblastoma, HIV and T cell-based cancer immunotherapy [6]. ZFNs targeting CCR5 to block CCR5–HIV interactions using a recombinant adenoviral vector was shown to result in the disruption of *CCR5* in >50% of transduced cells, both in model cell lines and primary human CD4 T cells. Treating hereditary disorders is another potential application of using ZFNs.

8.2.4 Challenges in ZFN

Several challenges remain to be addressed before we see the translation of ZFN from the laboratory to our daily life. Engineering the ZFNs is challenging, it is time consuming and an expensive technique. Genotoxicity or oncogenicity may be caused by potential off-target introduction of DSBs [7]. Appropriate delivery of ZFNs through safe and efficient means into the target cell still remains to be resolved.

However, ZFN mediated gene editing is an innovative application which removed several constraints on genetic modification that previously thought to be impossible or not achievable.

8.3 Transcription Activator-Like Effectors Nuclease (TALENs)

As an alternative to ZFNs, transcription activator-like effector nuclease (TALENs) emerged as a new editing tool. TALENs are a chimera of transcription activator-like effectors (TALEs) and restriction enzymes, predominantly *Fok*1 nuclease. The TALE proteins act as programmable DNA search engines while the *Fok*1 are the molecular scissors cutting the DNA target site.

8.3.1 Structure of Transcription Activator-Like Effectors (TALEs)

TALEs are T3SS effectors secreted by some β- and γ-proteobacteria. TALE proteins from plant pathogen Xanthomonas are most extensively studied. Once injected in host cells, these proteins reach the nucleus and manipulate promoter sequences of susceptibility (S) genes mainly interfering with nutrient transporters, host defense signaling and water-soaking regulation.

The TALE protein has a T3SS secretion signal sequence at the amino terminal. At the carboxyl terminal, an acidic activation domain, nuclear localization signals and transcription factor IIA site where the plant transcription factors can bind are found. The N-terminal and C-terminal region is separated by a signature DNA-binding central repeat region (CRR) comprising of 33–35 amino acid-long tandem repeats [8]. The repeats are mostly identical but vary primarily at amino acids 12 and 13 (termed the repeat-variable di-residue or RVD) and define the binding specificity of TALEs. Fig. 8.3 gives a schematic representation of the structure of TALE.

Crystal structure of TALEs shows that it has alpha helix structure. The helical structure permits TALEs to be in loose conformation with DNA during the search

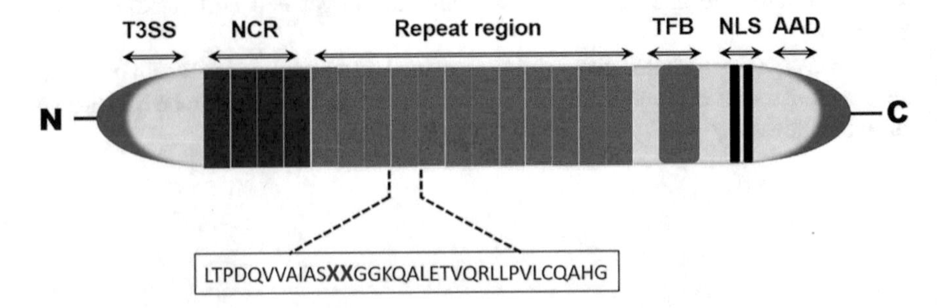

Fig. 8.3 Schematic of a representative TALE structure. TALE is composed of an N-terminal translocation domain (TD), a central repeat region, two nuclear localization signals (NLS) and activation domain (AD) in the C-terminal region. Each repeat variable di-residue (RVD) has 34 amino acids with two variable amino acids highlighted in red

Domain				
RVD	NI	NG	NK	HD
Nucleotide	A	T	G	C

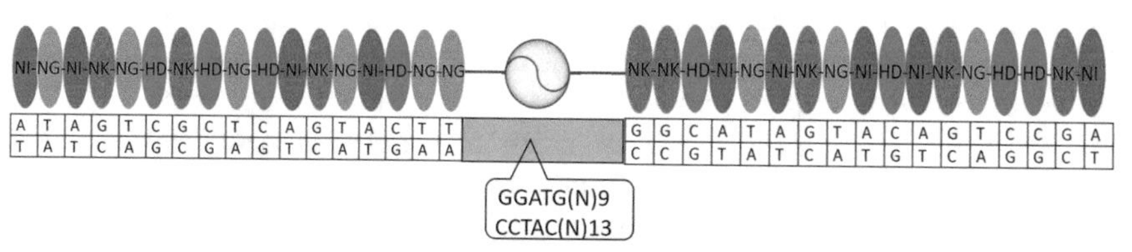

Fig. 8.4 General overview of TALEN process. Designed TALENs containing NI, NG, NK and HD recognize A, T, G and C are shown with blue, orange, green and purple respectively. The left and right TALENs recognize binding sites separated by spacer region. Dimerized *Fok*I recognizes the spacer region between two TALE sites and cuts DNA to generate DSBs

process facilitating rapid target location and then converting into condensed form to increase binding affinity upon entering the recognition site.

The binding of RVD to a distinct single nucleotide in the 5′ to 3′ orientation is stabilized by the asparagine or a histidine at position 12, while the residue at 13th position uniquely identifies a nucleotide through specific contacts. Elucidative analysis of the RVD has made it possible to map preferential binding of NN, NG, HD, and NI toward G/A, T, C, and A respectively. The "code-like" modular mechanism between amino acid sequence of RVD and individual DNA nucleotides enables custom designing of TALEs with desired specificities toward any target DNA.

The suites of transcription factors offer flexible DNA-binding domains (DBDs) which can be tethered with catalytic domain from *Fok1* restriction enzyme to create TALENs with sequence precision that bind and cleave DNA at pre-selected sites. Like ZFN, a pair of TALENs binds to opposite DNA strands in tail-to-tail orientations allowing the *Fok*I domain dimerization and cleavage of target DNA within the two binding sites (Fig. 8.4).

Several variants of TALENs developed and utilized for genome editing [9, 10] are discussed below.

8.3.2 *Variants of TALEN*

The advances in customization of TALENs have helped in proficiently fusing TALEs with several other restriction endonuclease such as PvuII, meganuclease I-TevI, meganucleases I-AniI, I-OnuI, DNA mismatch repair protein MutH, etc.

8.3.2.1 Methylation-Specific Manipulation

The variants TALE repeat N* is obtained by eliminating glycine from the 13th position (from NG) in the RVD modules. Due to absence of glycine residue, the RVD loop does not penetrate deep into the DNA major groove thereby avoiding interaction with methyl groups present in 5mC. TALE repeat N* couple with TALE Q* and TALE R* help TALEN models to discriminate between methylation status of cytosine nucleotides C, 5mC, and 5mhC, respectively, making it a methylation-dependent gene manipulating system.

8.3.2.2 Point Mutation

DNA point mutation can be achieved by installing a TALEN pair with cytidine deaminase (DddA) which facilitates C-G to T-A. Like *FokI* enzyme, DddA can be split into two functional halves DddA-N and DddA-C and coupled to two oppositely binding TALEs.

8.3.2.3 DNA Insertion

TALE carrying piggyBac transposases permits transposition of insert DNA at predetermined sites within the genome. The experiments related to utilization of TALE piggyBac fusion are in the optimization stage to obtain maximum insertion frequency.

8.3.2.4 TALERs

DNA invertase Gin as well as tyrosine recombinase Flp are fused to the N-terminal of TALE generating TALE recombinases (TALERs). Thus, TALEs are partnered with site-specific recombinases.

8.3.2.5 TALE Transcriptome Modifiers

TALE proteins fused with repressors such as Sid4, Kruppel-associated box (KRAB), EAR-(SRDX) has been developed to repress the transcription of a target gene. TALE DNA-binding domain coupled with DNA demethylase such as TET1 or methylases such as DNMT3A or DNMT3L can manipulate DNA methylation status, and thus result in activation or repression of target genes respectively. Likewise, TALE fusion with histone methyltransferases (KYP, TgSET8, NUE), deacetylases (hdac8, RPD3, Sir2a, and Sin3a) and demethylase (LSD1) are available to reveal the function of histone modifications on gene expression.

8.3.2.6 MitoTALEN

Along with genomic DNA, TALEN fused with mitochondrial targeting signal (MTS) are used to efficiently manipulate mitochondrial DNA. MitoTALEN successfully treated Leber's hereditary optic neuropathy, ataxia, neurogenic muscle fatigue, and retinal pigmentosa caused by mutation in human mitochondrial DNA.

8.3.3 Applications of TALENs

TALENs have been used to generate animal models for several diseases [11]. The LDL (low density lipoprotein) receptor in pigs has been inactivated by TALENs in pigs which serves as a model for hypercholesterolemia. Rabbit models for arteriosclerosis by knock-in of human apoAII have been successfully created. Human apolipoprotein A-II reduces atherosclerosis in knock-in rabbits. The human transcription factor NKX2–5 that plays an important role in cardiac formation and development has been targeted by TALENs in human hPSCs to better understand cardiac differentiation. The method can be used to generate Sp110 gene Knock-in resulting in greater resistance to tuberculosis in cattle. TALEN gene editing supplemented with template DNA facilitated gene knockout of locus resulting in a hornless phenotype thereby giving rise to hornless offspring. In humans, TALEN-edited CAR-T cells could effectively destroy leukemia causing tumor cells.

In plants, TALENs have been used to generate resistance against bacteria like *Xanthomonas* in rice. TALEN based targeting of multiple copies of mlo1 in wheat made it resistant to infection by powdery mildew, thereby limiting the need for pesticides. In sugarcane, targeting gene copies and alleles of the caffeic acid O-methyltransferase (COMT) resulted in 19.7% reduction of lignin and a 44% increase in saccharification making it a better-quality substrate for sugarcane-derived biofuel production. In soybean, mutation of two fatty acid desaturase genes reduced production of unhealthy trans fats by shifting the fatty acid profile from linoleic acid to oleic acid [12].

8.3.4 Challenges Associated with TALENs

TALENs exhibit a great promise to help and facilitate genetic modification in plants, animal and cell lines mainly due to its simple design and high success rate. However, the shortcoming of TALENs [7] is the size of the construct which is generally around kilobases, making it challenging to package and deliver it to model organisms.

8.4 RNA Interference

Fire and Mello awed the whole scientific community by injecting a double-stranded RNA (dsRNA) in *C. elegans* that could silence any gene sequence and produce associated phenotypes. This led to the discovery of RNA interference (RNAi). RNA interference (RNAi) is a neat and easy mechanism directed to suppress expression of genes by using a small RNA to complement the target sequence [13].

8.4.1 RNA Interference Mechanism

The RNAi machinery can be adapted in laboratory by introduction of synthetic small RNAs in the cells (Fig. 8.5). The most common small RNAs used are short interfering RNAs (siRNAs) or short hairpin (shRNAs). This small RNA is loaded into the RNA induced silencing complex (RISC) which in turn degrades the target mRNA which is complementary to the small RNA. There are some important steps in RNAi mediated gene editing, The dsRNA in the cytoplasm is processed by DICER, a RNAase III endoribonuclease generating around 21 nucleotide long siRNA. With help of dsRNA binding proteins, the RNAi-DICER subunit transfers the small dsRNA to Argonaute. In the next step Argonaute binds to one strand of the duplex and displaces the other strand. This whole complex is termed RISC. The completely formed RISC surveys for long RNAs and pairs with the single stranded guide RNA bound to Argounaute, which then degrades the target RNA through it RNAase-H activity. By this method mRNA and protein expression levels can be reduced.

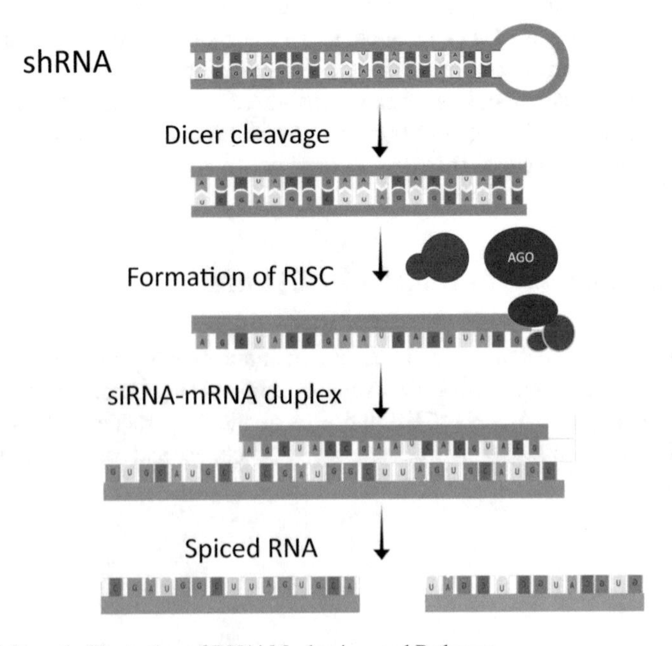

Fig. 8.5 Schematic Illustration of RNAi Mechanism and Pathway

8.4.2 Applications of RNA Interference

Manipulating the disease associated genes using RNAi as therapy has been of interest in the recent years. RNAi-based intervention has been explored in the field of cancer, autoimmune diseases, dominant genetic disorders, and viral infections [14]. It is one of the important tools for functional genomics. It is also one of the major mechanisms for creating animal models to understand the pathophysiology caused due to differential expression of certain proteins.

8.4.3 Challenges in RNA Interference

It has been studied that siRNAs can induce silencing of off-target mRNAs which have limited sequence complementarity through interactions with 3′UTR. A single siRNA can repress many transcripts if there are sequence imperfections or errors in the RNA [14]. Hence designing and validating the RNAi constructs becomes very important. There are several online tools that help construct optimization like si-Fi which is a BOWTIE-based sequence similarity search tool for putative siRNA constructs. It uses probability calculation of local target-site accessibility, and thermodynamics- as well as sequence-based prediction for strand selection.

Another major challenge in siRNA is the delivery of these molecules in the desired cell, as siRNAs do not cross the cellular membrane due to their size and negative charge. In the in vitro model, the delivery of siRNAs is mainly by cationic liposome-based methods. But in vivo, the rapid liver clearance and challenges in targeting specific tissue is inevitable. Modifying the RNA backbone using 2′F, 2′O-Me, and 2′H substitutions show to increase serum stability.

Taking caution to design accurate siRNA constructs, RNAi is the most easy and practical technology to achieve silencing of genes/proteins in mammalian somatic cells as the machinery is already present, it does not need any prior genetic modification of target cells required.

8.5 CRISPR-Cas9 System

Emmanuelle Charpentier and Jennifer Doudna received the Nobel prize in chemistry for pioneering the revolutionary Clustered Regularly Interspaced Palindromic Repeats (CRISPR) technology. The CRISPR-Cas9 genome editing technology is one of the most efficient platforms for genome editing in eukaryotic cells. This was first identified by Japanese researchers in *E. coli* genome. Further research showed that CRISPRs were involved in gene regulation or DNA repair and CRISPR-associated *cas* genes encoded proteins with helicase and nuclease domains [15].

8.5.1 CRISPR-Cas9 Gene Editing Mechanism

CRISPR-Cas9 system is a countermeasure adapted by prokaryotes to defend themself from infecting bacteriophages. It also acts as a memory bank allowing the bacteria to gain immunity which is analogous to the acquired immune system in vertebrates.

The system includes a Cas9 restriction enzyme that chops the invading foreign DNA into fragments using RuvC and HNH domains. CRISPR represents the collection of these fragments (20 bp) of foreign DNA incorporated into the bacterial host genome. The collected fragments also called protospacers are species specific and occur only once on CRISPR locus. During infection if a bacterium is attacked by virus or phage whose protospacer fragment is already present in the bacterium CRISPR DNA locus, the spacer is transcribed into pre-CRISPR RNA linked with Cas9 to form crRNA and activated upon addition of tracrRNA. The activated complex scans viral DNA and upon finding a match, the Cas9 latches onto the viral DNA and makes target-specific double-stranded DNA break, thus acting as a weapon preventing the virus from taking over the cell. Along with complementary match between viral and protospacer, Cas9 also requires NGG/NAG PAM (*Protospacer Adjacent Motif*) sequence to make a cut in viral DNA. As PAM is exclusively present in invading DNA and not in host DNA, the bacterial genomic DNA is protected from getting self-digested by its own Cas9 enzymes.

Along with Cas9 nucleases there is Cas12/Cpf1 family which has homologous RuvC domain but lacks HNH endonuclease domain making it compact as compared to Cas9. Additionally due to absence of HNH domain Cas12 generates staggered ends and depends on only crRNA rather than the two RNA (tracrRNA and crRNA) for cleavage. Unlike Cas9, Cas12 recognizes a different PAM site TTN/TTTN/TTTV (N is A/T/C/G; V is A/C/G) thereby expands the reach of CRISPR to edit regions of high AT content such as introns and promoters or AT-rich genomes.

As shown in Fig. 8.6, the class 2 CRISPR-Cas systems are the most preferred system for gene editing due to their simplicity and ease, in which the spacer sgRNA placed in the CRISPR direct the system to the target DNA and Cas9 controls the spacer function.

8.5.2 Variants of CRISPR-Cas9 System

Researchers have extensively explored the CRISPR-Cas9 system, creating an easily adaptable tool to target any genomic sequence by just customizing the guide RNA [16]. The availability of web-based tools such as chop chop, casfinder, flycrispr, ecrispir, crisprera made it easy to design and generate guide RNA allowing recruitment of Cas9 to any specific genomic loci, thus making it one of the most popular genome editing technologies.

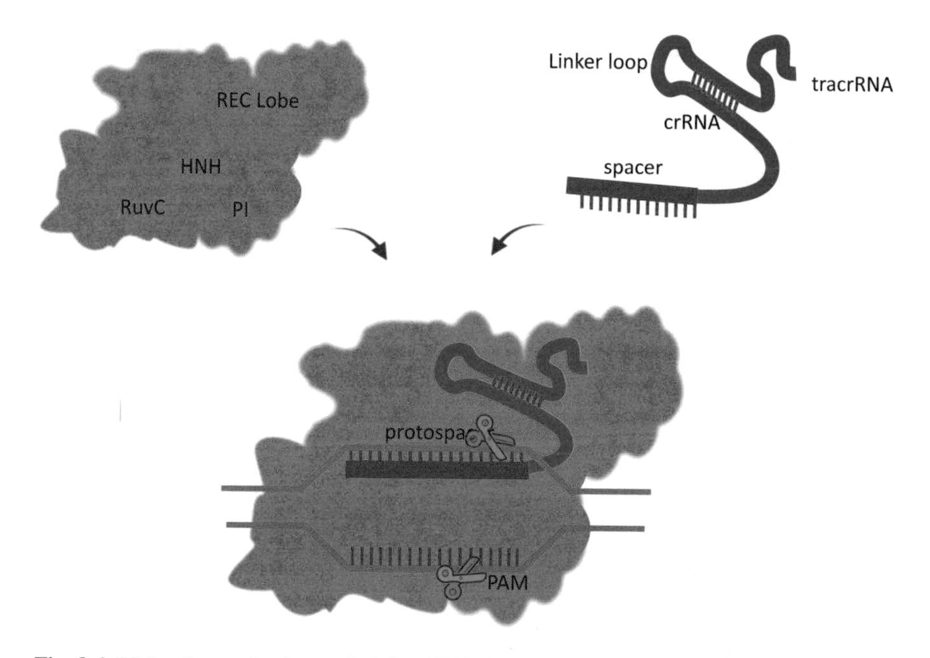

Fig. 8.6 Molecular mechanism underlying CRISPR-Cas9 genome editing: The sgRNA is composed of tracrRNA, crRNA and user-defined spacer complementary to target sequence. The co-expression of Cas9 and sgRNA leads to formation of ribonucleoprotein complexes capable of binding DNA. The annealing of spacer and target DNA along with PAM recognition serves as a signal to mediated DNA cleavage within the target (indicated by a pair of scissors)

8.5.2.1 Precision in Specific Gene Editing

Cas9 nickase, eSpCas9(1.1), SpCas9-HF1, HypaCas9, evoCas9, and xCas9 3.7 are variants of wild-type Cas9 obtained by mutating amino acids which are crucial in interactions between Cas9 and DNA strand. The mutation weakens the interactions between the enzyme and the substrate, allowing generation of DSB only when a strong complementary match is encountered, hence dramatically increasing the target specificity and preventing off-target editing.

The precision of gene editing is also achieved by manipulating the choice of DNA repair system employed to repair CRISPR/Cas9 induced breaks so as to favor the homology-directed repair HDR over the error prone DNA repair mechanism.

8.5.2.2 Precision in Base Editing

CRISPR-Cas9, although reputed for knocking out genes, can modify DNA by transitions and transversions of a single base with utmost precision. To fine-tune the system as a base editor, CRISPR with a catalytically impaired Cas9 nuclease was designed so that it can correct errors at single DNA nucleotides without making double-stranded breaks. For this, a variant of Cas9 with point mutation at D10A and H840A in the RuvC and HNH domain respectively was generated to obtain Cas9 "nickase." Cas9 nickase can generate nick on DNA strand but not cleave the phosphate backbone, only when dual nickase CRISPR system targeting opposite DNA strands are used simultaneously a double-strand breaks can be generated.

8.5.2.3 CRISPR in Gene Modulation

CRISPR interference (CRISPRi) and CRISPR activation (CRISPRa) are variants of CRISPR with nuclease dead Cas9 (dCas9) fused with gene activators or repressors. The dCas9 molecule has point mutations at D10A and H840A and hence, its catalytic activity is lost. The intact guide RNA navigates the genome and binds to the target region. Even though the dCas9 cannot cleave target DNA, it deposits the cargo effector molecule near regulator region resulting in RNA-directed transcriptional control of downstream genes.

In CRISPRi, loss-of-function of gene is due to repressor domains like KRAB (Krüppel associated box) or Mxi1 fused to dCas9. Similarly, CRISPRa cause gain-of-function due to transcriptional activators such as VP64, p65, SunTag. Use of CRISPR not only cause fewer off-target regulatory effects but also allows simultaneous repression/activation of multiple genes and is applicable to both coding and noncoding genes.

For irreversible and permanent gene modulation which can be inherited by daughter cells CRISPR–dCas9-based epigenetic editing has gain tremendous popularity. Inactive dCas enzymes fused with epigenetic modifiers like DNA methylases MQ1, histone acetyltransferases p300, histone demethylase TET1 or LSD1, and deacetylases are available and can be targeted to modify epigenetic status of specific locations within the genome.

8.5.2.4 CRISPR Fluorophores Imaging

dCas9 tagged with fluorescent probes have repurposed the CRISPR system into a customizable sequence-specific DNA labeler. Various colors are imparted to the CRISPR-Cas duo by using different fluorescent proteins offering site-specific multiplexed labeling in living cells under mild conditions. The fluorescence moieties are made integral part of CRISPR-Cas by different approaches such as tagging halo conjugates at C terminal of dCas9, tailing gRNAs with aptamers capable of binding fluorescent labeled RBPs, inserting fluorescent tagged hairpins (MS2, PP7, or boxB) into stem loops of the gRNA, etc.

Box 8.1

The computational and metagenomic studies recently identified a new Type VI CRISPR system Cas13 which is involved in protecting the bacteria from pathogens with RNA as genetic material. The diverse Cas13 family contains subtypes such as Cas13a (formerly C2c2), Cas13b, Cas13c, and Cas13d. Similar to Cas9, Cas13 use crRNA as guide RNA (60–64 bp), a short uracil-rich hairpin structure in crRNA allows the complex formation between Cas13 and guide RNA and facilitates target binding and cleavage. The target specificity is encoded by spacer sequence (28–30 nucleotides) that is complementary to the target RNA transcript. As compared to Cas9, Cas13 is smaller in size, cannot recognize PAM sequence as well as does not have RuvC and HNH domains. However Cas13a recognizes a single nucleotide motif "protospacer flanking site" (PFS) adjacent to 3′ end of spacer sequence and has HEPN domains essential for RNA cleavage, making Cas13 a RNA-guided RNA-interfering system which allows it to target sequences in RNA rather than DNA. This property of programmable binding and cleavage of RNA makes Cas13 a potential tool for influencing gene expression without altering genome sequence

To cleave the target RNA, Cas13 first has to find a match between crRNA and target RNA Cas13 then undergoes series of conformational changes to achieve enzymatically "active" state and only then chop the target RNA. After cutting the target sequence Cas13 does not revert to inactive states but continues to degrade surrounding RNA irrespective of their sequence. This collateral activity exhibited by Cas13 upon target recognition has been harnessed to cleave nearby fluorescent reporters making it a sensitive and precision diagnostic tool for example Specific High Sensitivity Enzymatic Reporter UnLOCKing (SHERLOCK).

In SHERLOCK, a reaction mix includes Cas13 carrying crRNA complementary to ssRNA of virus or pathogen along with quenched fluorescent ssRNA reporter molecules. If the given sample has virus or pathogen RNA, Cas13 recognizes and binds target RNA leading to activation of the RNAase domain. The activated Cas13 cleaves the target RNA as well as surrounding ssRNA reporter molecules producing a quantifiable signal that indicates the presence of virus or pathogen. The signal can be enhanced by amplification of template followed by T7 transcription to convert amplified DNA to RNA for subsequent detection by Cas13. SHERLOCK platform has been reported to successfully detect strains of Zika virus, genotype human DNA, and identify tumor mutations within cell-free genomic DNA. Similarly, to SHERLOCK, DNA specific CRISPR based diagnostic platform termed DNA Endonuclease Targeted CRISPR *Trans* Reporter (DETECTR) has also been developed. DETECTR is based on natural DNase activity of the Cas12a's to cleave nonspecific (*trans*) ssDNA. When Cas12a-cRNA base pairs with the dsDNA of interest, the DNase activity of Cas12a is initiated which not only cleave target DNA but also nonspecific *trans*-ssDNA in its vicinity, including the ssDNA-FQ reporter generating a quantifiable fluorescent signal. DETECTR is specific enough to accurately detect even the slight variation in HPV, HPV16, and HPV18. Another Cas12-based method HOLMES (for one-hour low-cost multipurpose highly efficient system) employs PCR and loop-mediated isothermal amplification (LAMP) as preamplification step to improve sensitivity of the CRISPR-based diagnostic tool.

As with Cas9, mutating key residues in the HEPN nuclease domain of Cas13 results in a "nuclease dead" Cas13a (dCas13a), that is capable of binding target RNA but lacks the ability to cleave the RNA target. Several Cas13-based systems such as REPAIR, RESCUE, TRM, PAMEC, etc. are developed as RNA editing tools.

REPAIR (RNA editing for programmable A to I replacement) is a fusion of adenosine deaminase (ADAR2) and dCas13b responsible for programmable conversion of adenosine to inosine in RNA. Another fusant, RESCUE (RNA Editing for Specific C-to-U Exchange) allows cytidine deamination and C to U exchange, while TRM (targeted RNA methylation) and PAMEC (photoactivatable RNA m^6A editing system using CRISPR-dCas13) edits RNA based on methylation status of adenine residues. The same approach is now adapted for fusion of dCas13 with splicing regulator (hnRNPA), ribo regulators, ribose modifiers which allows to program RNA stability, its location as well as its usage (translation).

8.5.3 *Applications of CRISPR-Cas9 Technique*

Though the CRISPR-Cas9 system is one of the most recent gene editing strategies, it is perhaps the most widely used one, with potential applications in almost every field of biology. In the beginning CRISPR was used to obtain gene knock outs in various cell types and organisms, but mutating Cas enzymes, blending enzymes with Cas, fusing gene regulators and epigenetic modifiers have extended application in precise genome editing, gene regulation, imaging DNA, and mapping chromosomes.

Potential application of this technique ranges from gene therapy to environmental sustainability. CRISPR is poised to reform the field of molecular medicine and gene therapy. Several researchers have proposed this technique to provide potential cure for diseases including neurodegenerative diseases, cancer, and hematological disorders. One of the very significant applications of CRISPR in molecular medicine is in chimeric antigen receptor (CAR) T cells which is a form of immunotherapy used to treat cancer. The use of this technique has shown to improve the safety and efficacy of the treatment as it can insert CAR at very specific sites in the genome. CRISPR-Cas system is a front runner in developing strategies to develop universal CAR-T cells which can be used to treat patients irrespective of their genotype. Recently, CRISPR has gained prominence in the field of molecular diagnostics. Molecular diagnostics engineered through the CRISPR-Cas system have been developed for infectious and genetic diseases. In agriculture, gene editing using CRISPR has been proposed for improvement of crop yield, development of pest resistant crops, development of drought resistant crops, increase the shelf life of food products, etc. Bioenergy is a key contributor toward development of a sustainable environment. CRISPR-Cas technique is being looked at for development of novel sources of bioenergy. In the laboratory, this has been used to knock out certain transcription factors that control lipid synthesis, thereby increasing the lipogenesis. This is being investigated as an important step toward generation of biodiesel.

The applications of the CRISPR-Cas system have endless capabilities in deciphering pathways, performing specific drug related studies to understand and help cure diseases. Controlling the expression, specificity, and expression of the CRISPR-CAS9 moieties and with better and accurate delivery methods this gene editing technology has huge potential in understanding cellular pathways.

8.5.4 Challenges Associated with CRISPR-Cas9 System

One of the major concerns for this approach of gene editing is that the Cas gene should be delivered into the cells for the expression of CAS9 protein [7]. This may activate the host immune system. Few studies show that lipid nanoparticle delivered RNA potentially activates toll-like receptors, followed by immune reactions. Hence, selecting the correct delivery vector and using a protein based CRISPR/CAS system could circumvent these issues.

8.6 Cre-loxP Method

Cre-loxP gene editing technology was discovered almost three decades ago by Nat. L Sternberg, where he identified the P1 gene product *Cre* (causes recombination) and its specific locus of crossover site named LoxP (locus of crossing (x) over, P1) [17]. It is one of the most common methods used for gene editing in vivo. Several animal models for research have been created using this technique.

8.6.1 Cre-loxP Gene Editing Mechanism

The Cre recombinase is a tyrosine recombinase enzyme (38 kDa) belonging to integrase family that catalyze the site-specific recombination event between two LoxP sites. Cre consists of 343 amino acids that form two distinct domains—the N terminal domain (NTD) and Catalytic domain (Cat) forming C shaped clamp that allows Cre protein to clench the DNA from opposite sides (Fig. 8.7). The first 129 residues constitute the amino terminal domain made up of 5 alpha helical (A-E). The helix A

Fig. 8.7 Domain map of Cre recombinase protein: Residues 1 to 129 represents N-terminal domain (black) connected by interdomain linker to the C-terminal catalytic domain (orange) stretching from 132 to 343 amino acid residues

along with helix E forms the tetramer structure of recombinase. Helix E is also involved in forming contact between the subunits while the LoxP binding property is attributed by helix B and D. The residues from 132–341 at carboxyl terminal harbors the active site of the enzyme. The C terminus has 9 distinct helices (F-N) which overall contributes to the protein stability and its docking onto the DNA. In the active site, the side chains of Arg 173, His 289, Arg 292 mark the phosphate at cleavage site for nucleophilic attack. Once the cleavage site is coordinated, Tyr 324 donates a pair of electrons to form a covalent 3′-phosphotyrosine bond with phosphate targeted at the cleavage site while Trp 315 form a hydrogen bond with this phosphate. This reaction cleaves the DNA and frees a 5′ hydroxyl group at the LoxP site.

The LoxP site consists of a pair of 13 bp palindromic sequences (ATAACTTCGTATA) separated by an 8 bp spacer sequence (ATGTATGC) making a total of 34 bp directional LoxP region. During Cre mediated recombination, a pair of LoxP sites flanks the gene of interest, each recognized by dimer of Cre protein. The flanking of the gene of interest by LoxP is referred to as "floxing." The position and alignment of lox sites determines whether the end result of recombination will lead to deletion, inversion, or translocation of a floxed DNA element.

8.6.2 Variants of Cre-loxP System

A variety of Cre recombinase systems are developed [18] with special features (Fig. 8.8) such as:

- *Inducible Cre-loxP system*: To achieve spatial and temporal control, an inducible Cre system has been generated. For example, the tamoxifen-inducible Cre system was made by fusion of the estrogen receptor harboring a mutated ligand binding domain (ER-LBD) with Cre protein known as CreER recombinase. This CreER along with tamoxifen (CreERT) present in the cytoplasm, binds to HSP90 (heat shock protein 90), the interaction is disrupted by tamoxifen or binding of similar synthetic steroids, causing nuclear translocation of CreERT in turn leading to Cre-loxP site interaction. The tetracycline (Tet) or doxycycline (Dox) inducible Cre systems are other cell specific and temporal inducible Cre systems.
- *Organ-specific promoters of Cre lines*: Several transgenic mouse lines have been developed where the activation/inactivation of the cre-loxP system is under the control of organ specific promoters. The specific promoters are available in the

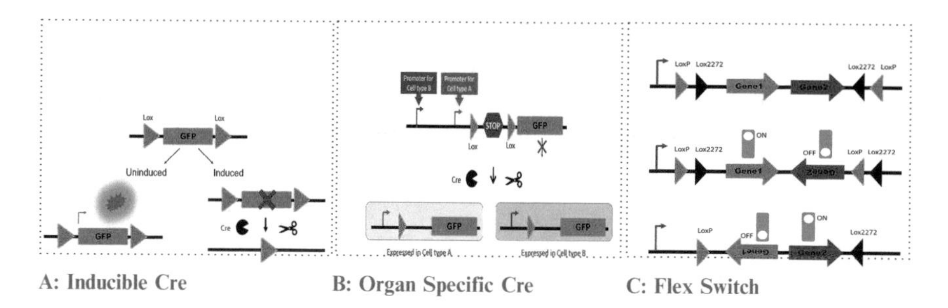

Fig. 8.8 Graphical illustration of variety of Cre recombinase plasmids

following databases: The Jackson laboratory (JAX cre-repository), National Center for Biotechnology Information (NCBI), and the International Mouse Genome Informatics (MGI). Specific promoters from organs from nervous system, immune system, musculoskeletal system, digestive system, and urogenital system have been used.

- *Split Cre*: To detect simultaneous expression of multiple genes, split Cre was developed by separating the NTD and Cat domain of the Cre recombinase into N-terminal and C-terminal fragments. The fragments are placed under the control of different promoters. Expression of one of the fragments is unable to cause DNA recombination. However, the co-expression of both N and C fragments in the single cell allows association of two fragments into functional enzymes. This new approach has been extensively useful in tracing neuron lineage, mapping brain circuits and gene targeting in terminally differentiated neuronal cells.
- *Flex Cre switch*: FLEX switches combine inversion and excision by employing two types of lox sites, e.g., loxP and lox2272. Here, two antiparallel ORFs are flanked by two pairs of LoxP variant recombination sites. As a result of this arrangement, Cre-dependent inversion of the ORFs activates one gene while repressing the other. FLEx switches are also known as DIO (Double-floxed Inverse Orientation) and DO (double-floxed orientation). Cre-on vectors are called DIO as they flip the gene of interest converting it from off state to ON state, while Cre-off vectors are called DO because a gene in the correct orientation is inverted, turning off its expression. This tool is extremely flexible and can be used for controlling multiple genes, inducing tissue specific gene expression and gene knock-in.

8.6.3 Applications of Cre-LoxP System

Cre LoxP is one of the most powerful editing tools to bring about genetic modification of animals. It is one of the most important tools used to create animal models required for a specific study. The tools have now become inevitable in creating models to study the molecular pathophysiology of diseases like cancer, neurodegenerative diseases, metabolic diseases, etc. [19]. It is also used in discovery and development of novel targeted therapy against diseases. Besides this, the toll can also be used to create models to study functional mutations, tissue specific role of proteins and effect of co-expression of different genes.

8.6.4 Limitation of Cre-LoxP and Advances

Although the Cre-loxP system does not need any cofactors, accessory proteins, or other template sequences, generating gene knockouts using this system are very time consuming and laborious involving continuous screening and selection. The application of this system in high throughput activities is limited. The Cre-loxP system may generate unpredictable mutants and may also trigger off-target effects on sites resembling LoxP [19].

Recently there have been efforts to combine the Cre-loxP system with an efficient direct cloning method called Cas12a-directed cloning (CAPTURE) that allows direct cloning of high GC content DNA sequences and very large genomic fragments around 113 kb in vivo [20]. Thus two powerful gene editing tools can be combined together to modulate genes for desired phenotypes.

8.7 Summary

Gene editing tools are one of the most recent milestones and the most advancing fields in recombinant DNA technology. A plethora of literature have reported the tools available for gene editing, advancement in methods and techniques to refine the applications of each tool available. Most of the recent advancements in the field of life sciences including gene therapy, understanding molecular basis of diseases, crop improvement and creating a sustainable environment can be attributed to the technical developments in the field of gene editing. Various tools like ZNF, TALEN, RNAi, CRIPSR, etc. are promising solutions to resolve the unmet needs that help improve the quality of life on earth. However, further knowledge of precision and tight regulation of genome editing would allow us to overcome the limitations associated with gene editing tools.

Table 8.1 Below summarizes the different aspects of gene editing covered in this chapter.

Self Assessment

Q1. If your experiment requires the creation of a mice model with expression of housekeeping gene ‘A’ repressed only in the liver, which method will you prefer?

A1. The preferred method will be gene editing using Cre-loxP technique. The Cre recombinase is a tyrosine recombinase enzyme (38 kDa) belonging to the inte-

Table 8.1 Summary of the gene editing methods covered in this chapter

	ZPN	TALENs	RNAi	CRISPR-cas9	Cre-loxP
Loss-of function mechanism	Transcription repression	Transcription repression	Post-translational RNA degradation	Frameshift DNA mutation	Gene deletion/inversion/ translocation
End result	Permanent knockdown	Permanent knockdown	Reversible knockdown	Permanent knockdown	Permanent gene modification
Advantages	Small protein size allows efficient packaging	Very high specificity with no requirement repeat links	Knockdown achieved in a dose dependent manner	Multiple targeting of genes is possible	Efficient and stable gene editing in animal models
Disadvantages	Low efficiency	Sensitive to DNA methylation	Off-target effects	Requirement of PAM sequence	Transgenic systems time consuming and expensive

grase family that catalyze the site-specific recombination event between two LoxP sites. The LoxP site consists of a pair of 13 bp palindromic sequences (ATAACTTCGTATA) separated by an 8 bp spacer sequence (ATGTATGC) making a total of 34 bp directional LoxP region. During Cre mediated recombination, a pair of LoxP sites flanks the gene of interest, each recognized by dimer of Cre protein. Using this technique, one can bring the editing of a particular gene under the control of organ specific promoters.

Q2. Mention any two methods by which the site specificity of CRISPR Cas gene editing can be improved

A2. Mutation in Cas9 gene weakens the interactions between the enzyme and the substrate, allowing generation of DSB only when a strong complementary match is encountered, hence dramatically increasing the target specificity and preventing off-target editing.

The precision of gene editing is also achieved by manipulating the choice of DNA repair system employed to repair CRISPR/Cas9 induced breaks so as to favor the homology-directed repair HDR over the error prone DNA repair mechanism.

References

1. Ranjha L, Howard SM, Cejka P. Main steps in DNA double-strand break repair: an introduction to homologous recombination and related processes. Chromosoma. 2018;127(2):187–214.
2. Urnov FD, Rebar EJ, Holmes MC, Zhang HS, Gregory PD. Genome editing with engineered zinc finger nucleases. Nat Rev Genet. 2010;11(9):636–46.
3. Miller JC, Holmes MC, Wang JB, Guschin DY, Lee YL, Rupniewski I, et al. An improved zinc-finger nuclease architecture for highly specific genome editing. Nat Biotechnol. 2007;25(7):778–85.
4. Novak S. Plant biotechnology applications of zinc finger technology. Methods Mol Biol. 2019;1864:295–310.
5. Chenouard V, Remy S, Tesson L, et al. Advances in genome editing and application to the generation of genetically modified rat models. Front Genet. 2021;12:615491.
6. Schwarze LI, Głów D, Sonntag T, et al. Optimisation of a TALE nuclease targeting the HIV co-receptor CCR5 for clinical application. Gene Ther. 2021;28:588–601.
7. Khan SH. Genome-editing technologies: concept, Pros, and Cons of Various Genome-Editing Techniques and Bioethical Concerns for Clinical Application. Mol Ther Nucleic Acids. 2019;16:326–34.
8. Ma AC, Chen Y, Blackburn PR, Ekker SC. TALEN-mediated mutagenesis and genome editing. Methods Mol Biol. 2016;1451:17–30.
9. Miller JC, Tan S, Qiao G, et al. A TALE nuclease architecture for efficient genome editing. Nat Biotechnol. 2011;29:143–8.
10. Carlson DF, Tan WF, Lillico SG, Stverakova D, Proudfoot C, Christian M, et al. Efficient TALEN-mediated gene knockout in livestock. P Natl Acad Sci USA. 2012;109(43):17382–7.
11. Mohr SE, Smith JA, Shamu CE, Neumuller RA, Perrimon N. RNAi screening comes of age: improved techniques and complementary approaches. Nat Rev Mol Cell Bio. 2014;15(9):591–600.
12. Setten RL, Rossi JJ, Han SP. The current state and future directions of RNAi-based therapeutics. Nat Rev Drug Discov. 2019;18:421–46.

13. Sigoillot FD, King RW. Vigilance and validation: keys to success in RNAi screening. ACS Chem Biol. 2011;6(1):47–60.
14. Ran FA, Hsu PD, Wright J, Agarwala V, Scott DA, Zhang F. Genome engineering using the CRISPR-Cas9 system. Nat Protoc. 2013;8(11):2281–308.
15. Haft DH, Selengut J, Mongodin EF, Nelson KE. A guild of 45 CRISPR-associated (Cas) protein families and multiple CRISPR/Cas subtypes exist in prokaryotic genomes. PLoS Comput Biol. 2005;1(6):474–83.
16. Sauer B, Henderson N. Site-specific DNA recombination in mammalian cells by the Cre recombinase of bacteriophage P1. Proc Natl Acad Sci U S A. 1988;85(14):5166–70.
17. McLellan MA, Rosenthal NA, Pinto AR. Cre-loxP-mediated recombination: general principles and experimental considerations. Curr Protoc Mouse Biol. 2017;7(1):1–12.
18. Tian X, Zhou B. Strategies for site-specific recombination with high efficiency and precise spatiotemporal resolution. J Biol Chem. 2021;296:100509.
19. Song AJ, Palmiter RD. Detecting and avoiding problems when using the Cre-lox system. Trends Genet. 2018;34(5):333–40.
20. Enghiad B, Huang CS, Guo F, Jiang GD, Wang B, Tabatabaei SK, et al. Cas12a-assisted precise targeted cloning using in vivo Cre-lox recombination. Nat Commun. 2021;12(1):1171.

Chapter 9
Applications of Recombinant DNA Technology

Madhura Vipra, Nayana Patil, and Aruna Sivaram

9.1 Introduction

In the previous chapters, you have read the basics of molecular cloning and recombinant DNA technology (RDT). One of the most versatile technologies, RDT has grown phenomenally over the last couple of decades due to better understanding of basic sciences, development of bioinformatics tools and overall technological advances. RDT as defined by the Encyclopedia Britannica is "the joining together of DNA molecules from different organisms and inserting it into a host organism to produce new genetic combinations that are of value to science, medicine, agriculture and industry." The advent of RDT brought about remarkable changes in the lives of mankind, mainly in the form of innovations and strategies to address three key concerns of human beings—health, food, and environment. Besides these, it also contributed toward increasing the yield and quality of industrial products like primary and secondary metabolites, enzymes, etc. Figure 9.1 gives an idea about various arenas where RDT plays an impact.

Despite the best of efforts by researchers and scientists, several unmet medical needs still persist. RDT has had a great impact in human healthcare and is playing a great role with some spectacular innovations leading to advanced healthcare with both preventive and therapeutic applications. With this technology it became easier

M. Vipra
Medvolt Tech Pvt. Ltd, Pune, Maharashtra, India
e-mail: madhura@athenasystech.com

N. Patil · A. Sivaram (✉)
School of Bioengineering Sciences & Research, MIT ADT University, Pune, Maharashtra, India
e-mail: nayana.patil@mituniversity.edu.in; aruna.sivaram@mituniversity.edu.in

N. Patil, A. Sivaram, *A Complete Guide to Gene Cloning: From Basic to Advanced*, Techniques in Life Science and Biomedicine for the Non-Expert,
https://doi.org/10.1007/978-3-030-96851-9_9

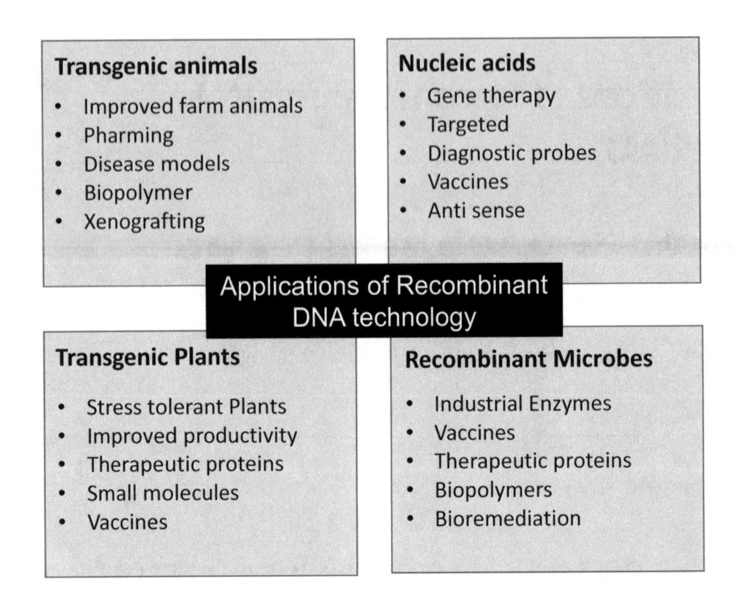

Fig. 9.1 Applications of RDT. RDT plays an important role in healthcare, sustainable environment, agriculture, and in industrial products. The contribution of RDT toward these areas have been significant and have made a remarkable impact

to produce engineered recombinant proteins in easily cultivated microorganisms or in vitro cultured cells leading to discovery of several new potential therapeutics and tools for diagnosis. Various therapeutic proteins like insulin, growth hormone, interferon, several antibiotics, monoclonal antibodies, etc. have been produced using RDT. Vaccines have played a very important role in shaping up human lives in terms of facing their ability to survive diseases which were earlier deadly and would wipe away populations. The deaths that occurred in Covid-19 pandemic are representative of how infectious diseases caused by bacteria and/or viruses can spread rapidly causing severe disease resulting in a very high death percentage. The entire field of genomics that includes diagnostics, clinical genomics, consumer genomics and Gene therapy is essentially evolved from gene mapping. rDNA technology helped mapping of the genome which has revolutionized the healthcare industry and human life. Human genome project, one of the greatest feats achieved by the human race was started on October 1, 1990 and completed in April 2003. Led by an international team of researchers, it resulted in sequencing and mapping of the genome of our species. This was the start of an entirely brand new, exciting, and possibly unending era of what is now known as Precision Biology as for the first-time nature's complete genetic blueprint for building a human being was mapped [1].

Pauling and his colleagues in 1949 introduced the concept of molecular diseases. This concept was based on their studies in sickle cell anemia where they found that the disease is caused by a change in a single amino acid. As more and more studies were carried out in this field, molecular diagnostic methods were developed. Last two decades have seen increasing use of molecular diagnostic methods. Such recently developed tests along with some new tests that are in the research phase will be pivotal in the delivery of safe and effective therapy for many diseases in the future.

Sustainable environment affects all living things on the earth alike. The challenges posed by pollutants and contaminants is increasing day by day, resulting in economical, ecological, and societal burden which may bring about an existential crisis for several species on the earth [2]. With increasing pollution and destruction to the environment comes the burden of climate change. Bioremediation and bioindicators are two very important developments in the field of environmental bioengineering which is likely to bring about a paradigm shift in the environment that we live in. The application of bioremediation is an evolving field of study in microbiology and environmental engineering. The biotechnological applications of microorganisms have viable potential of solving the significant dangers.

Primary metabolites are mostly intermediates and/or end products of anabolic metabolism. Other primary metabolites from catabolic metabolism (e.g., citric acid, acetic acid, and ethanol) are related to energy production and substrate utilization, again essential for growth. Industrially, these microbial metabolites are very important, some examples being amino acids, nucleotides, vitamins, solvents, and organic acids. Advent of recombinant DNA technology in protein engineering led to the production of more efficient processes [3]. In agriculture, development of genetically modified crops with an objective to improve both yield and resistance to plant pests and herbicides has gained a significant degree of public and government acceptance and is already being practised in a commercial context in several countries.

This chapter covers myriad applications of RDT with a greater focus on few of imminent importance.

9.2 Medicine and Healthcare

RDT has helped in overcoming several challenges encountered in the field of healthcare like development of novel therapies, vaccines and tools for diagnosis. Protein engineering became an exciting new field with tremendous potential and has helped in development of several novel technologies relevant to healthcare.

In the subsequent sections we will discuss some of the interesting applications of RDT with respect to the healthcare sector.

9.2.1 Role of RDT in Vaccine Development

The first recombinant vaccine, approved in 1986, was produced by inserting a gene fragment from the hepatitis B virus into yeast [4]. Other examples of infectious diseases against which immunity is achieved through vaccination include polio, tetanus, diphtheria, measles, and hepatitis, etc. The vaccines exert their action by activating the immune system against the pathogens. The success rate is more against infections caused by pathogens that have low antigenic variability as protection against them can be achieved with antibody-mediated immunity. However, pathogens that rapidly mutate pose a challenge as immunity generated by vaccines may not work on the new variant and may need frequent immunization, like in Influenza virus. Although attenuated viruses are used for vaccination, production of vaccines through the rDNA method has gained lots of attention. These vaccines are considered to have a better safety profile as they do not contain the entire virus, instead carrying only specific proteins or genetic codes for these proteins.

We will have a detailed look at the recent Covid-19 Pandemic and see how RDT has helped face the devastating effects (Fig. 9.2). There were enhanced efforts by scientists from prestigious pharmaceuticals resulting in rapid release of vaccines within a short span of around 6–7 months. The most available vaccine in India, Covishield is ChAdOx1-S, which is a recombinant vaccine developed by AstraZeneca and Oxford University. In this vaccine, adenovirus, which causes very mild or asymptomatic infections in humans is used as the vector to deliver the SARS-CoV-2 spike protein. This spike protein causes immunological response resulting in immunity against SARS-CoV-2. Adenoviral based vectors Ad5 and Ad26 are used in another vaccine based on RDT called Sputnik V. Another approach used for Covid-19 is that of RNA vaccines. Here the mRNA carrying the genetic sequence for the protein (spike protein of SARS-CoV-2) are enclosed usually in

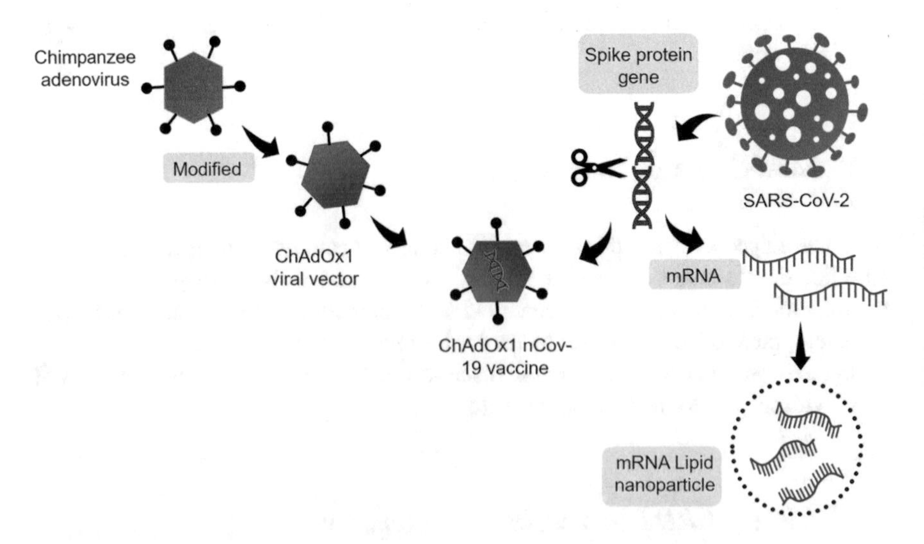

Fig. 9.2 Illustration of recombinant vaccine approaches for COVID-19. In DNA vaccines a chimpanzee adenovirus is modified into a low seroprevalence ChAdOx1 viral vector. The vector is engineered to carry an optimized full-length SARS-CoV-2 spike protein gene. In the RNA vaccine, mRNA strand encoding a S protein gene of SARS-CoV-2 is produced. The mRNA is coated and encapsulated in a lipid nanoparticle

nanoparticles which resemble cell membranes and delivered into the human cell. Thereafter, the host uses its machinery to make spike protein, which is recognized by the host's system, triggering an immune response.

9.2.2 Role of RDT in Therapeutic Protein

Several therapeutics have been developed using RDT. Humulin, or Human insulin was one of the very first products of RDT which found widespread acceptance and received approval from FDA. This is an era for novel biological entities and biosimilars, which have proved to be game changers in targeted therapy due to their improved specificity and efficacy. Given below is a list of some of the engineered recombinant proteins that are in actual use for humans.

- *Synthesis of Human Insulin*: Insulin, a hormone involved in glucose metabolism is very important for human health. It is used by patients with diabetes. Human genes for insulin production have been incorporated into bacterial DNA and such genetically engineered bacteria are used for large-scale production of insulin [5, 6].
- *Synthesis of Human Growth Hormone*: Human growth hormone was produced by inserting DNA coding for human growth hormone into a plasmid that was implanted in *E. coli* bacteria. HaeIII, a type of restriction enzyme which acts at restriction sites in the 3′ noncoding region and at the 23rd codon in complementary DNA for human growth hormone, was used to produce a DNA fragment of 551 base pairs which includes coding sequences for amino acids 24–191 of HGH [7].
- *Synthesis of Monoclonal antibodies*: Cesar Milstein and George Kohler won the Nobel Prize in 1984 for inventing a technique that produced the first monoclonal antibodies [8] Tocilizumab (TCZ), is a recombinant humanized anti-interleukin-6 receptor (IL-6R) monoclonal antibody which was developed for use in the treatment of rheumatoid arthritis, systemic juvenile idiopathic arthritis (sJIA) and polyarticular juvenile idiopathic arthritis (pJIA). Since then several monoclonal antibodies have been generated and have been approved by FDA for treatment of various diseases.
- *Synthesis of Interferon*: Interferons are glycoproteins produced as a part of immune response by cells which are infected with viruses. These proteins have antiviral properties and act as the first line of defence. However, natural interferon is produced in very small quantities hence very costly. rDNA technology now can produce interferon at a much cheaper rate. E.g., Interferon alpha is used to treat lymphoma and myelogenous leukemia.
- *Synthesis of Antibiotics*: The application of RDT was based on the development of genetic transformation systems for β-lactam-producing organisms and cloning of biosynthetic genes. Yield improvements through metabolic engineering have been demonstrated for a number of systems. For example, the production of

cephalosporin C by *Cephalosporium acremonium* was increased by 15% by overexpressing the cefEF gene [9]. *Penicillium chrysogenum* strains showed that increase in penicillin could result by introducing extra copies of biosynthetic genes and by increasing copy number and high transcription levels of the whole cluster. Production was increased by overexpressing the gene encoding phenyl-acetic acid-activating CoA ligase from *Pseudomonas putida*. Further 30-fold increase in penicillin production was achieved by overexpression of gene acvA in *Aspergillus nidulans*, by replacing the normal promoter with the ethanol dehydrogenase promoter [10].

- *Synthesis of antitumor agents*: The first recombinant drugs approved in 1986, were interferon-α2a (IFN-α2a) and interferon-α2b (IFN-α2b) for the treatment of hairy cell leukemia, Kaposi sarcoma, chronic myeloid leukemia, malignant melanoma and follicular lymphoma. FDA approved antitumor agents from microorganisms include actinomycin D (dactinomycin), anthracyclines glycopeptolides, etc. Ofatumumab, Bevacizumab, Alemtuzumab, Ipilimumab, Trastuzumab are some of the recombinant products approved by FDA [11].

9.2.3 Molecular Diagnostics

The molecular diagnostic tests are now available for a large number of medical diseases such as infectious disease, oncology, immune function tests, coagulation, and for pharmacogenomics. The molecular diagnostics have moved into the next-generation technology with the advent of synthetic biology [12].

Below are some of the applications of RDT in diagnostics.

- *Prenatal diagnostics*: RDT will permit screening for many more genetic disorders as a part of prenatal diagnosis. These techniques help in identification of mutations and genes related to a particular disease, thus enabling quicker treatment.
- *Pharmacogenomics/Clinical Genomics/Consumer Genomics*: A highly specialized field that evolved over last two decades, where in SNPs (Single nucleotide polymorphism), a single nucleotide change in genome or combinations of SNPs can be detected and can be associated with prevalence of disease in specific ethnic population and also help predict response of patients to certain drugs based on their SNP profile. For example, the enzyme CYP2C19 metabolises drugs, such as the anti-clotting agent Clopidogrel, into their active forms. Some patients may possess mutations (polymorphisms) in 2C19 gene which can lead to poor metabolism of the drug leading to possibility that drug is not effective. Physicians can test for these polymorphisms and find out whether the drugs will be fully effective for that patient [13]. Another example of it is Methylene tetrahydrofolate reductase (MTHFR), which is an enzyme that is necessary to convert inactive Folic acid into the active form of folate 5-methyltetrahydrofolate (5-MTHF) that does all the folate functions in our body. MTHFR polymorphism affects up

to 10% of the population which means those many people cannot process Folic acid. Folate is very important for several body functions specially related to DNA, and if excess folic acid is not converted into folate can cause some effects which are not good for body such as suppression of immune function, anemia, cardiovascular disease. B12 synthesis and pathways are also linked to MTHFR. Thus, knowing the genetic variant present in a person doctors can help design his folic acid requirement actually helping him.

- *Infectious Diseases*: Molecular methods have a clear advantage over traditional microbiological testing in many ways including speed, accuracy etc. One of the greatest success stories has been the development of diagnostics for rapid detection of *M. tuberculosis* by Genexpert from clinical specimens. The Xpert MTB/RIF is a cartridge-based nucleic acid amplification test (NAAT) for simultaneous rapid tuberculosis diagnosis and rapid antibiotic sensitivity test [14]. It is an automated diagnostic test that can identify *Mycobacterium tuberculosis* (MTB) DNA and resistance to rifampicin (RIF). In December 2010, the World Health Organization (WHO) endorsed the Xpert MTB/RIF for use in tuberculosis (TB) endemic countries.

9.2.4 Gene Therapy

Gene therapy started gaining momentum in the early 2000s with the approval of Gendicine, for head and neck cancer in China. Later Neovasculgen was approved in Russia for peripheral artery disease in 2011. Gene therapy is the current focus of researchers as it is the most advanced therapeutic use of DNA technology in health-care (Fig. 9.3). This essentially involves repairing or reconstructing defective genetic material responsible for the ailment. Few potential therapeutic interventions, include tumors (e.g., head and neck cancers), new blood vessels in the heart to alleviate heart attacks and to stop HIV-replication in patients with AIDS, but applications could be vast. Recent successes in genetic medicine have laid the foundation for next-generation technologies.

Two different approaches are often adopted for gene therapy. In an ex vivo approach, cells are detached, genetically modified, and transplanted back into a patient. For example, bone marrow from the patient is removed and grown in laboratory conditions, is transfected with the desired gene and returned to the patient. In the in vivo approach, the genetic material is directly transferred into. Viral vectors are used for gene therapy as they rely on the host machinery for their replication and transcription. Researchers around the globe have made significant contributions to overcome the challenges associated with viral vectors.

Nonviral systems like liposome, DNA–polymer conjugates and naked DNA are also used to transfect the target cell. These methods present certain advantages such as ease in large-scale production and low host immunogenicity. However, nonviral methods are found to be nonspecific with lower success rate as it produces lower levels of transfection and gene expression, and thus lower therapeutic efficacy.

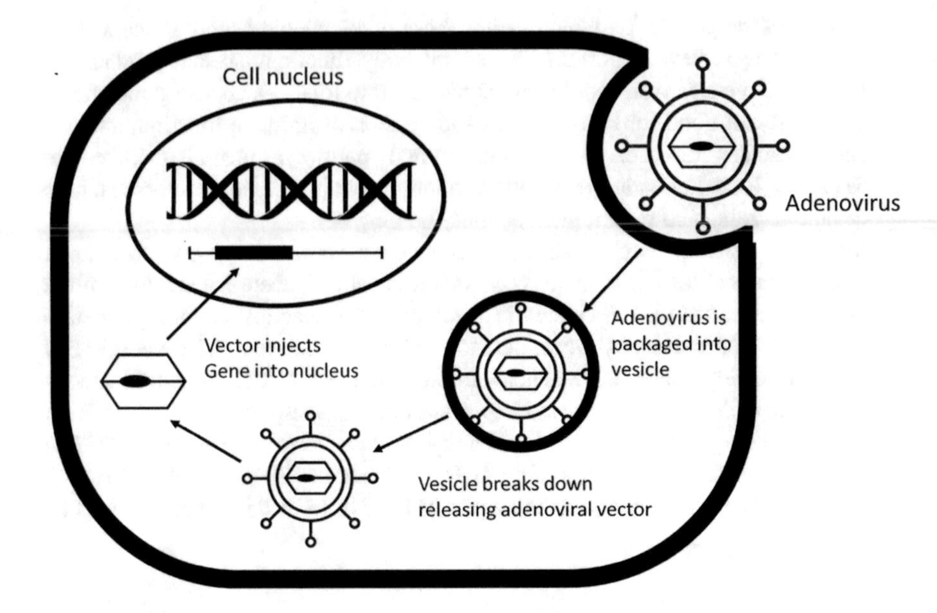

Fig. 9.3 Schematic representation of transfer of gene of interest into the host cell using adenoviral vectors. The gene of interest is cloned into the vector, which enters the cell and releases the genetic material into it

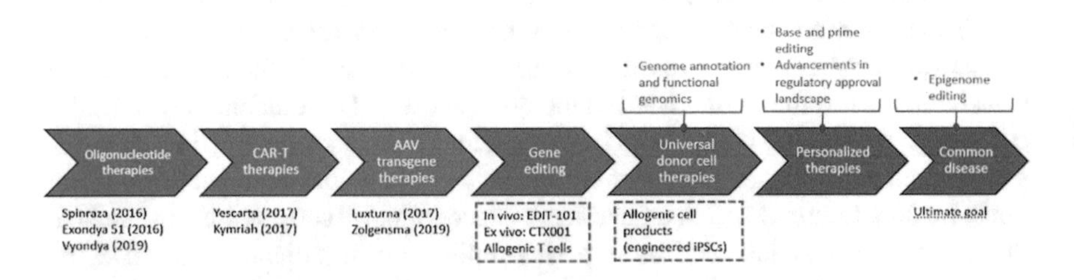

Fig. 9.4 Timeline depicting milestones of general gene therapy development

Newer technologies that offer the promise of solving these problems have been developed. Zinc finger nucleases, TALENs, and CRISPR are the common systems used for gene editing [15]. One of the most well-known gene therapy methods is CAR-T, used for cancer treatment. A general timeline of development made in gene therapy is represented in Fig. 9.4.

9.3 Sustainable Environment

There are different categories of bioremediation methods like Enhanced Natural Attenuation and Bio-stimulation for removing pollution and reducing the deleterious effect of environmental contaminants. Bioaugmentation is a technique that enhances the capacity of a contaminated biotope for decontamination by deliberately introducing precisely competent strains or groups of microorganisms which locally alter the existing biome. It is employed where conventional bioremediation methods have failed. The basic utility of such an intervention is that the metabolic capacities of existing microbiomes in contaminated biota will be augmented by an externally engineered exogenous genetic diversity, thus leading to a larger repository of viable biodegradation reactions. Genetically Engineered Microorganisms (GEMs) with the capability of increased degradative activity have been developed [13, 16]. A number of biotic and abiotic factors affect bioaugmentation. The most relevant abiotic factors are temperature, pH, organic matter content, nutrient content, and soil type. The significant biotic factors are inter-microbial competition for resources, antagonistic microbial interactions, and killing by protozoa.

9.3.1 Treatment of Municipal Wastewater

Bioaugmentation is commonly used in municipal wastewater treatment for activated sludge bioreactors. Activated sludge systems are generally based on microorganisms like bacteria, protozoa, nematodes, rotifers, and fungi capable of degrading organic matter, but in combination with GEMs can be re-engineered to adequately perform treatment of wastewater. These consortia usually comprise of microorganisms such as *Bacillus licheniformis, Bacillus thuringiensis, Paenibacillus polymyxa, Bacillus stearothermophilus, Penicillium sp., Aspergillus sp., Flavobacterium,* etc. Such technologies are important more so in urban set-up and have enabled efficient wastewater treatment with local treatment plants capable of treating large quantities of wastewater and thus preventing pollution of local water bodies.

9.3.2 Treatment of Groundwater

Bioaugmentation is used to remove contaminants like chlorinated ethenes from groundwater. It allows complete degradation of contaminants due to microbial activity of in-situ microorganisms into ethylene and chloride, which are nontoxic. Several literature reports cite evidence wherein bioaugmentation has been used to degrade other pollutants like chloroethanes, chloromethanes, and methyl tert-butyl ether.

9.3.3 Heavy Metal Removal

The application of RDT for removal of heavy metal contamination from the environment has generated a lot of scientific and commercial interest. Removal of chromium metal from industrial effluents has been reported using *Alcaligenes eutrophus*

AE104 (pEBZ141). A RDT based system was developed wherein a photosynthetic bacterium, *Rhodopseudomonas palustris* was engineered to express genes responsible for removal of mercury contamination from wastewater.

9.4 Manufacture of Industrial Products

Metabolic engineering is a key technique in the industry these days. It is used to enhance formation of products through certain biochemical reactions. RDT is being widely used for metabolic engineering of industry products. This is achieved through genetic engineering techniques like site directed mutagenesis where specific and intentional changes are made to the DNA sequence of a gene and thus in gene products.

9.4.1 Role of RDT in Production of Primary Metabolite

Corynebacterium, Brevibacterium, and *Serratia* are routinely used for the commercial production of amino acids. Plasmid vector systems for cloning in *Corynebacterium glutamicum* were established leading to increased amino acid production by the bacterium and related strains [17]. A multi-fold increase in the yield of L-lysine was seen in mutants of *C. glutamicum*. As compared to the wild type strains, these mutants showed an increase in the yield by over 20 fold [18]. This was achieved through several approaches like deletion of certain genes like PEP carboxykinase, glucose-6-phosphate dehydrogenase, or by introducing mutations which decreased the enzymatic activity of citrate synthase. Overexpression of certain enzymes like pyruvate carboxylase or DAP dehydrogenase also helped in improving the yield.

Similar observations have been made with Biotin. A difference in the yield by around 10,000 times was seen when biotin operon was cloned in a high copy number plasmid. This high copy number plasmid helped in increased expression of the genes in the operon, thereby increasing the yield. Classical mutation and screening/selection techniques as well as genetic engineering has resulted in strains of *E. coli* and *C. glutamicum* yielding high levels of acetate, pyruvate, succinate, and lactate.

Ethanol, an alcohol, is a very important industrial chemical. *E. coli* was converted into a good ethanol producer using RDT. In the genetically engineered strain, the majority of the products of fermentation was ethanol. Sorbitol, also called D-glucitol, is produced by hydrogenation of D-glucose through enzymatic reactions. It is a sweetening agent with uses in various industries like food and pharmaceutics. Around 500,000 tons of sorbitol is manufactured across the industries per year on a global level. Genetic engineering of *Lactobacillus plantarum* by increasing the expression of the two sorbitol-6-phosphate dehydrogenase genes (srlD1 and srlD2) resulted in high sorbitol production. Other examples of industrial products where RDT is used include, nucleosides, carotenoids, glucosamine, acetone, etc.

9.4.2 Role of RDT in Production of Secondary Metabolites and Enzymes

Secondary metabolites do not play a significant role in normal growth and development of an organism. The type of microorganism and the specific environment in which it grows adds to diversity of secondary metabolites. They have various applications in food, medicine, and other industries. Using RDT, the production of various secondary metabolites has been increased [19]. Enzymes are the catalysts synthesized by living cells. They are complex protein molecules that bring about different important chemical reactions acting as a catalyst in all life forms. For industrial enzyme production microorganisms are used. Enzyme engineering, involves production, isolation, and purification of new enzymes with chosen functions. Enzymes have wide range of applications and are used in food industry such as food production, food processing and preservation, in soaps and washing powders, in disinfectants, in biofuel, in dyeing and textile manufacture, in leather industry, in pulp and paper industry, several medical applications, in agriculture, in ceramic industry, in crude oil extra in environment management and of course in scientific research to name a few.

9.5 Agriculture

RDT in plants is usually performed using *Agrobacterium*-mediated methods. RDT has been used globally to improve crops and forests. It has become one of the invaluable tools to improve the features of plants and their products. There are a number of features which scientists are trying to improvise and optimize through the application of these technologies such as resistance to disease, pests, and other forms of stress, nutritional balance and protein content; and herbicide resistance.

9.5.1 Nutritional Enhancement of Crops: Production of Golden Rice

Deficiency of vitamin A (formed from Vitamin A) is a major problem in countries where rice the staple food. Food primarily composed of rice does not contain vitamin A, though it does have a precursor molecule which can be converted into β-carotene by certain enzymes present in some bacteria and plants. In a genetically engineered strain of rice called Golden rice, β-carotene is produced from the precursor molecules. Conversion of C20 into phytoene is a rate limiting step in the synthesis of β-carotene. This step is catalyzed by the enzyme phytoene synthase, white is absent in white rice. Introduction of the phytoene synthase gene into rice enabled the production of golden rice, which is rich in beta carotene. Besides the gene for phytoene synthase, scientists have also introduced genes for Phytoene desaturase and Lycopene beta-cyclase into rice genome to activate synthesis of β-carotene.

9.5.2 Production of Slow Ripening Fruits

First genetically modified (GM) food licensed for human consumption was the Flavr Savr tomato (Fig. 9.5). The ripening process in these tomatoes is slowed down by inhibition of an enzyme polygalacturonase, which is involved in the cellular processes leading to the formation of ethylene. Ethylene acts as a hormone which accelerates fruit ripening.

The genetically engineered enzymatic inhibition leads to the inhibition of ethylene which results in the delayed ripening of tomatoes and acquired resistance to rotting. This has benefits both from a harvesting perspective as well as from a logistics and storage point of view.

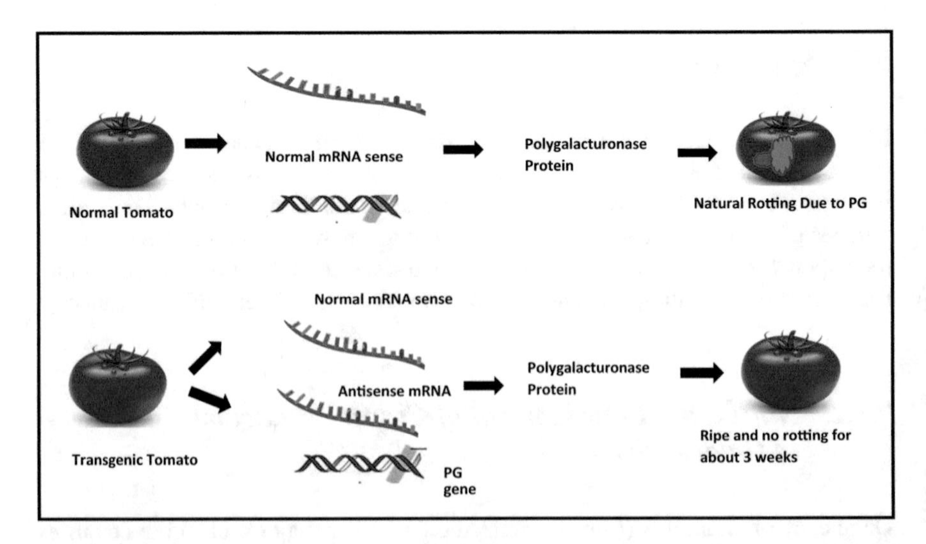

Fig. 9.5 RDT in Food. Innovations in agriculture and the food industry have helped in extending the shelf-life of products and also in streamlining the process of crop harvesting

9.5.3 Resistance to Herbicides and Pests

Bacillus thuringiensis (*Bt*) is a gram-positive soil bacterium which, on sporulation, produces toxins with crystalline morphology coded for by cry genes which makes them resistant to certain pests. To produce insect-resistant transgenic plants, the genetic sequence coding for the toxin is separated from the bacterium, and with suitable promoter and terminal sequences, engineered recombinant DNA is constructed and introduced into the plant using *Agrobacterium tumefaciens* as the vector which contains a plasmid (Ti) that carries the desired gene.

The *Bt* toxin encoded by the genes cry I Ac and cry II Ab control Bollworms which damage the crops like cotton to a large extent, while cry I Ab toxin controls Corn Borer. Bt toxins are produced in inactive form but are converted to their active form in the alkaline pH environment of the insect gut. This interferes with the digestion process of pests eventually leading to their death. *Bt* toxins bind to the inside of the insect gut and damage its epithelium so that it cannot absorb digested food and thus the insects starve to death. Transgenic cotton, tomato, potato, and tobacco plants resistant to various pests have been created. However, concerns have been also raised on the long term effect of modified crops on humans and environment [20].

9.6 Summary

RDT plays a very important role in almost all walks of life—majorly in addressing the global challenges related to health, agriculture, environment, and industrial products. Today, the world has reached a situation where one has to depend on RDT for one's daily existence and survival. It has by-and-large become one of the inevitable tools to maintain quality of life and sustainable environment and for production of sufficient food for the global population. More advanced technologies of RDT will help it expand its horizon further, bringing about a paradigm shift in our daily lives.

Self Assessment

Q1. How does RDT help in improving industrial products?

A1. RDT helps in improving the yield of primary metabolites like carotenoids and riboflavin. It also has improved production of ethanol and sorbitol. An increased yield of several secondary metabolites and enzymes have also been seen when appropriate RDT are used for production.

Q2. How did RDT help during the Covid-19 pandemic?

A2. Several vaccines against SARS CoV-2 virus were produced using RDT. Covishield is ChAdOx1-S, which is a recombinant vaccine developed by AstraZeneca and Oxford University. In this vaccine, adenovirus, which causes very mild or asymptomatic infections in humans is used as the vector to deliver the SARS-CoV-2 spike protein. This spike protein causes immunological response resulting in immunity against Sars-CoV-2. The second vaccine based on RDT is Sputnik V, which uses two human adenoviruses Ad5 and Ad26. The adenoviruses used in Covid-19 vaccines are non-replicating, which means that the virus vector in the vaccine does not multiply after entering the human body. Another approach used for Covid-19 is that of RNA vaccines. Here the mRNA carrying the genetic sequence for the protein (spike protein of SARS-CoV-2) are enclosed usually in nanoparticles which resemble cell membranes and delivered into the human cell. Thereafter, the host uses its machinery to make spike protein, which is recognized by the host's system, triggering an immune response.

References

1. Gates AJ, Gysi DM, Kellis M, Barabás A-L. A wealth of discovery built on the human genome project — by the numbers. Nature. 2021;590:212–5.
2. Joutey NT, Bahafid W, Sayel H, El Ghachtouli N. Biodegradation: involved microorganisms and genetically engineered microorganisms. Biodegrad Life Sci. 2013;1:289–320.
3. Sanchez S, Demain AL. Metabolic regulation and overproduction of primary metabolites. Microb Biotechnol. 2008;1(4):283–319.
4. Glick BR, Patten CL. Molecular biotechnology: principles and applications of recombinant DNA. John Wiley & Sons; 2017.
5. Bristow AF. Recombinant-DNA-derived insulin analogues as potentially useful therapeutic agents. Trends Biotechnol. 1993;11(7):301–5.
6. Bryan PN. Protein engineering of subtilisin. Biochim Biophys Acta. 2000;1543(2):203–22.
7. Cronin MJ. Pioneering recombinant growth hormone manufacturing: pounds produced per mile of height. J Pediatr. 1997;131(2):S5–7.
8. Liu JH. The history of monoclonal antibody development – Progress, remaining challenges and future innovations. Ann Med Surg (Lond). 2014;3(4):113–6.
9. Skatrud PL, Tietz AJ, Ingolia TD, Cantwell CA, Fisher DL, Chapman JL, Queener SW. Use of recombinant DNA to improve production of cephalosporin C by Cephalosporium acremonium. Bio/Technology. 1989;7(5):477–85.
10. Kennedy J, Turner G. γ-(L-α-aminoadipyl)-L-cysteinyl-D-valine synthetase is a rate limiting enzyme for penicillin production in aspergillus nidulans. Mol Gen Genet. 1996;253:189–97.
11. Grilo AL, Mantalaris A. The increasingly human and profitable monoclonal antibody market. Trends Biotechnol. 2019;37(1):9–16.
12. Tan X, Letendre JH, Collins JJ, Wong WW. Synthetic biology in the clinic: engineering vaccines, diagnostics, and therapeutics. Cell. 2021;184(4):881–98.
13. Ozyigit II, Can H, Dogan I. Phytoremediation using genetically engineered plants to remove metals: a review. Environ Chem Lett. 2021;19:669–98.
14. Park M, Kon OM. Use of Xpert MTB/RIF and Xpert ultra in extrapulmonary tuberculosis. Expert Rev Anti-Infect Ther. 2021;19(1):65–77.
15. Belete TM. The current status of gene therapy for the treatment of cancer. Biologics. 2021;15:67–77.
16. Pant G, Garlapati D, Agrawal U, Prasuna RG, Mathimani T, Pugazhendhi A. Biological approaches practised using genetically engineered microbes for a sustainable environment: a review. J Hazard Mater. 2021;405:124631.

17. Blombach B, Schreiner ME, Moch M, Oldiges M, Eikmanns BJ. Effect of pyruvate dehydrogenase complex deficiency on L-lysine production with Corynebacterium glutamicum. Appl Microbiol Biotechnol. 2007;76(3):615–23.
18. Radmacher E, Eggeling L. The three tricarboxylate synthase activities of Corynebacterium glutamicum and increase of L-lysine synthesis. Appl Microbiol Biotechnol. 2007;76(3):587–95.
19. Romsdahl J, Wang CC. Recent advances in the genome mining of aspergillus secondary metabolites (covering 2012-2018). Medchemcomm. 2019;10(6):840–66.
20. Belousova ME, Malovichko YV, Shikov AE, Nizhnikov AA, Antonets KS. Dissecting the environmental consequences of bacillus thuringiensis application for natural ecosystems. Toxins (Basel). 2021;13(5):355.

Chapter 10
Ethical and Safety Concerns of Recombinant DNA Technology

Richa Sharma, Nayana Patil, and Aruna Sivaram

10.1 Introduction

The year 1982 marked the approval of the first recombinant human insulin—Humulin by the US FDA and there was no looking back after that. Approximately 275 recombinant drugs/vaccines have been approved around the world for human ailments so far. Genetically modified plants with resistance to diseases, capable of tolerating physical or chemical stress, and flexibility to adapt along with improved production dominate the agriculture sector.

Needless to say, the advances in the recombinant DNA technology had started bothering the scientists, for the unprecedented events that the technology may lead to. Considering the impact of the research around the sciences, the scientific fraternity, in the historic ASILOMAR conference in mid-1970s, laid the foundation for the NIH guidelines for the recombinant DNA technology and the outcomes thereafter. The first version of the NIH guideline was released in 1976 and has been regularly updated. Similarly, in the field of agriculture, combined efforts taken by Inter-American Institute for Cooperation on Agriculture (IICA) and Inter-American System have laid down standards, policies, and guidelines for the implementation, preservation, and regulation of the technologies. The First Meeting of the Regional Directing Council of the Regional Biotechnology Program, UNDP/UNESCO/ONUDI 1987, established "the Pan American Health Organization (PAHO) in order to consult the

R. Sharma
Labcorp Scientific Services & Solutions Pvt. Ltd., Pune, Maharashtra, India

N. Patil · A. Sivaram (✉)
School of Bioengineering Sciences & Research, MIT ADT University, Pune, Maharashtra, India
e-mail: nayana.patil@mituniversity.edu.in; aruna.sivaram@mituniversity.edu.in

N. Patil, A. Sivaram, *A Complete Guide to Gene Cloning: From Basic to Advanced*, Techniques in Life Science and Biomedicine for the Non-Expert,
https://doi.org/10.1007/978-3-030-96851-9_10

subject experts, policy maker and more over bring public awareness about the biosafety of the recent technologies and product generated using them" [1].

10.2 Questionable Knowledge Dissemination

In 1999, 18-year-old Jesse Gelsinger died while participating in a gene therapy trial at the University of Pennsylvania. This led to the criticism for the institution for its failure to disclose crucial information on informed consent documents, relaxing criteria for accepting volunteers, and enrolling volunteers who were ineligible. So we see, each innovation has more than one public opinion, ethical committee advice, and national policy framework. The questionable knowledge sharing pertaining to the drug candidate, its efficacy, adverse events, or the patient disclosure to the regulatory body, still persists.

Furthermore, confidentiality while sharing the genetic data of human subjects is a major concern. The easy access to technology and data brings in further concerns to this graving issue. Thankfully, regulations such as the Health Insurance Portability and Accountability Act, 1996 (HIPAA) are in place [2, 3].

The HIPAA and GINA are the genetic privacy acts that have been strategically put to place after the induction of the Human Genome Project. The implications on the development of sciences or restrictions posed by these acts still remain controversial [4].

10.3 Uncertainties about Misuse of Technology

A lab generated microorganism or a mutated species might have nuisance value once it is out in the world. A therapeutically inserted vector with a cancer gene might cause cancer in the exposed personnel or those that happen to be exposed to it. A probability of the genetic information being stolen with/without permission also exists. The story of creation of the HeLa cell line for the involuntarily donated cells of Henrietta Lacks is an example of this. The family got to know about the cell donation only after her death and they never received due compensation, although a lot of people reaped benefits out of it.

Manipulation of the agricultural genes for better yields, improved efficacy against pests and infections have led to emergence of many bioengineered species of crops, which might lead to unintended events. Another example of such manipulation was the *Bt* cotton (*Bacillus thuringiensis*) controversy, the genetic impact of *Bt* cotton and the pollen movement on biodiversity, environment and wild species had not emerged clearly and the seeds were rolled out on the fields.

10.4 Potential Risk of Engineered Organisms

The outburst of the pandemic Coronavirus infections added fuel to the fire of the potential risk of engineered organisms. The controversy of coronavirus mutations being of natural origin or genetically introduced is still ongoing, and there has been more politics and science into it than before. GMO, if released in the environment, can lead to ecological disasters and affect animal welfare. Issues with inadequate handling and storage or containment of biohazard material at laboratories or industrial scale may lead to accidental release of the engineered organisms into the environment [5, 6].

10.4.1 Forensic Sciences

When we consider the application of RDT in forensic sciences and the possibilities for its misuse or abuse; the reliability, validity, and confidentiality of the tests still remains questionable. For this Committee on DNA Technology in Forensic Science deals with using DNA technology in forensic science.

10.4.2 Transgenic Animals

The idea of genetic engineering brought a boom in the pharmaceutical industry and the transgenic animals that were once thought of, now became the reality. These transgenic animals have been widely used for conducting various preclinical studies, designing specific disease models (e.g., mouse with knockout genes for specific cancers/genetic diseases/metabolic ailments). They facilitate better understanding of the disease process and drug targeting for the same. Transgenic animals have also been greatly used for the production of recombinant forms of biological proteins, hormones (e.g., insulin, growth hormone). These animals have also been used to generate organs for human use (xenografts). The issue of graft rejection still remains for many of these strategies. They have also been used as pets. Transgenic animals have also found ways into our kitchens. Transgenic animals designed with manipulation of a specific gene, however, might cause allergic reactions to the human who consumes it. Thus, these are high risk ones. Laboratory animals, although have a risk of accidental exposure, have a lower potential risk than the farm animals [7].

Based on the risk involved, the impact on the ecosystem, researchers have set metrics around the use of transgenic animals. One strategy of classification has been shared below as class I, class II, and class III.

"Class 1—Laboratory animals: essentially used to get knowledge and not for direct profit, show frequent unpredictable side effects, used in limited numbers: possible tolerance towards suffering.

Class 2—Animals used as sources for organs or pharmaceuticals: directly used for human health, may generate high profit, may suffer from known and reproducible deleterious side effects, used in limited numbers: tolerance towards suffering on a case-by-case basis.

Class 3—Farm animals and pets: not strictly required in most cases for human survival, may generate profit, may show known and reproducible deleterious side effects, used in large numbers: no tolerance towards suffering".

On similar lines, transgenic plants too, raise similar alarms. They might lead to allergic reactions, toxicity, or antinutrient effects [8].

10.5 Regulations and Guidelines on Biosafety of GMOs

As evident from the above examples there is an incremental risk associated with GMO raising the sense of concern among scientists and policy makers. The solution lies in systematic and structured policies that are implemented for ensuring safety of GMOs and products thereof. On the basis of scientific knowledge and consultation with experts, academicians, and stakeholders, a regulatory framework of guidelines is followed worldwide. The global policies, regulations, and guidelines have the mandate to monitor rDNA research, projects involving hazardous model systems and release of GMO products. The regulatory bodies also evaluate biosafety of organisms, appropriate containment facilities for handling risk group organisms and waste disposal.

Even though the guidelines are followed globally, it may be updated, optimized, and evolved by each country to suit their needs. Table 10.1 consolidates globally established Recombinant DNA Advisory or Monitoring Committees [9].

With the view to cope up with rapid advancement in biotechnology research and for decisive measures for biosafety, India has established regulations and guidelines on biosafety, biocontainment, and responsible use of Recombinant DNA Research. As mandated in the Rules, 1989 of Environment (Protection) Act, 1986, Ministry of Science and Technology, India, has recently updated and released new regulations and guidelines in 2017 [10]. There are six established committees authorized for successful implementation of the rules right from local surveillance to national and international levels.

1. Recombinant DNA Advisory Committee (RDAC).
2. Institutional Biosafety Committee (IBSC).
3. Review Committee on Genetic Manipulation (RCGM).
4. Genetic Engineering Appraisal Committee (GEAC).
5. State Biotechnology Coordination Committee (SBCC).
6. District Level Committee (DLC).

Together the committee establishes standard for

Table 10.1 The regulation agencies by geographical regions

Region	Regulators
Europe	European Food Safety Authority
USA	United States Department of Agriculture (USDA) Animal and Plant Health Inspection Service (APHIS) Environmental protection agency (EPA) Food and Drug Administration (FDA)
Latin Africa	National Biosafety Management Agency (NBMA) Academy of Sciences of South Africa (ASSAf),
India	Recombinant DNA Advisory Committee (RDAC) Institutional Biosafety Committee (IBSC) Review Committee on Genetic Manipulation (RCGM) Genetic Engineering Appraisal Committee (GEAC) State Biotechnology Coordination Committee (SBCC) District Level Committee (DLC)
China	Ministry of Agriculture of China Administration of Quality Supervision Inspection and Quarantine Office of Agricultural Genetic Engineering Biosafety Administration (OAGEBA)
England	Department for environment food and rural affairs
Australia	Office of the Gene Technology Regulator Therapeutic Goods Administration Food Standards Australia New Zealand
Canada	Plant with novel traits PNT, Canadian Food Inspection Agency (CFIA)
Argentina	Commission on Agricultural Biotechnology (CON ABIA) Food and Agriculture Organization of the United Nations (FAO)
Brazil	National Biosafety Technical Commission (environmental and food safety) Council of Ministers (commercial and economical issues)
New Zealand	Hazardous Sub[1]stances and New Organisms Act 1996 (HSNO) Food Standards Australia New Zealand (FSANZ)
Japan	Ministry of the Environment (MOE) Ministry of Agriculture, Forestry and Fisheries (MAFF) Ministry of Health, Labor and Welfare (MHLW)
Mexico	Mexican Department of Health

- Safe handling of GMO and pathogenic microorganisms.
- Classification of organism into risk groups, appropriate containment facilities, and certification of facilities.
- Regulate the handling, production, and exchange of GMO, pathogenic microorganisms, and products.
- Accessibility and information of safe usage and handling of GMO.
- Public awareness about the containment strategies followed in India.

In order to measure the threat and benefits associated with GMO, appropriate guidelines, regulations, and legislation should be adopted by the global scientific community. The success of GMOs in healthcare, bioremediation, and agriculture must be judged based on the risk to human health and environment.

10.6 Summary

Laws for regulation of outcomes of RDT have been designed by different regulatory bodies, and they are specific to the region and its ethnicity, views, etc. The consequences of the product developed via RDT should be well accessed before it is made public. Potential risks associated with each of the application of RDT should be considered and dealt with as per the regulations.

Self Assessment

Q1. Give any two risk factors associated with engineered organisms

A1. Answers: Genetically altered organisms, if released in the environment can increase human suffering, lead to ecological disasters and affect animal welfare. Issues with inadequate handling and storage or that in containment of biotechnological material in laboratories and industrial plants might lead to accidental release of the engineered organisms into the environment. Identify and list the successful projects based on RDT that have enhanced the lives of human beings.

Q2. Give any two ways of overcoming the risk associated with GMOs

A2. Answers: The solution lies in systematic and structured policies that are implemented for ensuring safety of GMOs and products thereof. On the basis of scientific knowledge and consultation with experts, academicians, and stakeholders, a regulatory framework of guidelines is followed worldwide. The global policies, regulations, and guidelines have the mandate to monitor rDNA research, projects involving hazardous model systems and release of GMO products. The regulatory bodies also evaluate biosafety of organisms, appropriate containment facilities for handling risk group organisms and waste disposal.

References

1. Bib Orton. Guidelines for the use and safety of genetic engineering techniques or recombinant DNA technology. IICA. Miscellaneous Publications Series DER/USA/88/001). 1988. ISSN 0534-5391.
2. Silverman E. The 5 most pressing ethical issues in biotech medicine. Biotechnol Healthc. 2004 Dec;1(6):41–6.
3. Nass SJ, Levit LA, Gostin LO. Beyond the HIPAA privacy rule: enhancing privacy, Improving Health through research. National Academies Press (US): Washington, D.C.; 2009.
4. Norrgard K. Protecting your genetic identity: GINA and HIPAA. Nat Educ. 2008;1:21.
5. Häyry M, Takala T. Biotechnology and the environment: from moral objections to ethical analyses. In: The proceedings of the twentieth world congress of philosophy, vol. 1; 1999. p. 169–78.
6. Hayry M, Häyry H, Chadwick R, Levitt M, Whitelegg M. Cultural and social objections to biotechnology: analysis of the arguments, with special reference to the views of young people. In: Ina report to the Commission of the European Communities. Centre for Professional Ethics; 1996.

7. Houdebine LM. Impacts of genetically modified animals on the ecosystem and human activities. Global Bioethics. 2014;25(1):3–18.
8. Dona A, Arvanitoyannis IS. Health risks of genetically modified foods. Crit Rev Food Sci Nutr. 2009;49(2):164–75.
9. Turnbull C, Lillemo M, Hvoslef-Eide TAK. Global regulation of genetically modified crops amid the gene edited crop boom – a review. Front Plant Sci. 2021;12:630396.
10. Barse B, Yazdani SS. India—GMOs/synthetic biology rules/regulations and biodiversity–a legal perspective from India. Cham: Springer; 2020. p. 559–76.

Index

N. Patil, A. Sivaram, *A Complete Guide to Gene Cloning: From Basic to Advanced*, Techniques in Life Science and Biomedicine for the Non-Expert,
https://doi.org/10.1007/978-3-030-96851-9

Printed in the United States
by Baker & Taylor Publisher Services